Shape Memory Alloy Valves

Alexander Czechowicz • Sven Langbein

Editors

Shape Memory Alloy Valves

Basics, Potentials, Design

 Springer

Editors
Alexander Czechowicz
FGW e.V.
Remscheid, Germany

Sven Langbein
Menden, Germany

ISBN 978-3-319-35445-3   ISBN 978-3-319-19081-5   (eBook)
DOI 10.1007/978-3-319-19081-5

Springer Cham Heidelberg New York Dordrecht London

Printed on acid-free paper

Springer International Publishing AG Switzerland is part of Springer Science+Business Media (www.springer.com)

Preface

Shape memory alloys (SMA) are multifunctional materials with remarkable characteristics for application in medical, industrial, and consumer products. On one hand they are suitable as superelastic materials for implants and medical devices; on the other hand they can be used as actuators using thermal or electrical energy to generate strokes and actuating forces.

Due to their benefits in industrial applications like miniaturization of drives, massive weight reduction, noiseless actuation, and intelligent intrinsic functions, they are interesting as substitution and superior concepts for actuators. Especially actuators for utilization in valve drives are in the focus of shape memory alloy applications, as so presented in this book.

Often research and development collaborations with industrial and academic partners focus on analogical questions about the application of these smart materials in drives. Often these drives have to be used in valves, as several famous applications have already proven the suitability for series.

Nevertheless, the calculation of displacements, lifetime, and dynamic properties are just three examples in the wide field of shape memory alloy actuator development which often are unclear to an unexperienced engineer.

Hence, the aim of this book is to introduce shape memory alloy technology, with focus on valve drive applications. While the basic chapters present the valve and SMA technology basically, the rear chapters discuss proceedings during product development.

The book is written for designers in machine-, valve-, automotive-, aerospace engineering, product developers, and product managers as well as for teachers and students at universities of science and mechanical engineering.

Remscheid, Germany
Menden, Germany
January 2015

Alexander Czechowicz
Sven Langbein

Contents

Contributors

Alexander Czechowicz Zentrum für Angewandte Formgedächtnistechnik, Forschungsgemeinschaft Werkzeuge und Werkstoffe e.V., Remscheid, Germany

Sven Langbein FG-INNOVATION GmbH, Bochum, Germany

Stefan Seelecke ZeMA – Zentrum für Mechatronik und Automatisierungstechnik gemeinnützige GmbH, Saarbrücken, Germany

Konstantin Lygin Ruhr-Universität Bochum, Bochum, Germany

Michael Hannig Bürkert Werke GmbH, Ingelfingen, Germany

Falk Höhne Lindner Armaturen GmbH, Chemnitz, Germany

Chapter 1
Introduction

Alexander Czechowicz and Sven Langbein

Shape memory alloys (SMA) are multifunctional materials with unique characteristics in the wide field of actuators. These materials can transform their form into an imprinted shape by thermal activation after a macroscopic deformation. In their high temperature phase, large reversible mechanical deformations can be reached. This transformation of the material phase allows the usage of SMA in actuating applications with remarkable benefits in the field of valve technology.

Due to their simple form, the complexity of SMA in applications is often underestimated. While an SMA actuator can look like a metallic wire, it still has multifunctional characteristics which have to be considered during the system development. This fact is the main hindrance for the assertion of SMA as valve drives on the market. Due to thermal balances, mechanical behavior and a complex effect-network, a shape memory wire actuator is a high-tech system. A major reason for errors during the development of such systems occurs often in the concept phase already. There critical application relevant parameters may be ignored or misjudged. On the other hand, conventional drive principles, like solenoids or bimetals, facilitate the development process by availability of development norms, guidelines, and experiences.

In the field of shape memory technology, many publications have been prepared on the material background and the material behavior itself. For an intensive study of the material characteristics and the modeling of the functions, the book edited by D. Lagoudas "Shape memory alloys" (Springer USA) is an excellent source of knowledge.

A. Czechowicz (✉)
Zentrum für Angewandte Formgedächtnistechnik, Forschungsgemeinschaft Werkzeuge und Werkstoffe e.V., Papenberger Straße 49, 42859 Remscheid, Germany
e-mail: czechowicz@fgw.de

S. Langbein
FG-INNOVATION GmbH, Universitätsstr. 142, 44799 Bochum, Germany
e-mail: langbein@gmx.com

A. Czechowicz, S. Langbein (eds.), *Shape Memory Alloy Valves*,
DOI 10.1007/978-3-319-19081-5_1

The following book concentrates on the product development from the strategic and methodical point of view. Hence, the aim of the book is to sensitize the user to the properties of the material, its potentials, and its application in valve systems.

Starting with an introduction in valve technology, the requirements of today's systems are presented by valves which have gained acceptance on the market compared to strategic ideas using shape memory alloy drives. The SMA technology is introduced by a brief explanation of the effects and by the description of SMA actuator applications with cross connections between the most relevant dynamic, electric, and mechanic parameters. A special attention lies also on the intrinsic sensor functions of the SMA material.

After the basic chapters, the book focuses on the application potentials in industrial markets and especially in valve technology. The potentials reveal not only examples for substitution of accepted valve drives, but also on new ideas and integrated functions which can be achieved only by utilization of SMA.

In the application chapters, the book presents calculation methods and guidelines for the development of thermal and electrical SMA valve drives. According to the methodical product development and for enhancement of the understanding, an exemplary pinch valve development is presented in Chap. 10. This development is done by guiding through an excerpt of the standardized German guideline VDI2248 for SMA actuator development which will be published in winter 2015.

Examples of SMA valve products which have appeared on the market verify the applicability of this unique technology for controlling fluid flows and justify the development challenges described in the precedent chapters. An outlook for future perspectives shows that further new potentials will be added to this technology field in the future.

Chapter 2
Valve Technology: State of the Art and System Design

Michael Hannig, Falk Höhne, and Sven Langbein

2.1 Introduction: Actuated Valves in Applications

Valves regulate, direct or control the flow of fluids by opening, closing, or partially obstructing various flow channels. In an open valve, fluids like gases, liquids, fluidized solids, or slurries, flow in a direction from higher pressure to lower pressure. Actuated valves use external energy to switch these flow channels by actuator systems. These actuators are driven by changes in pressure, electrical energy, temperature, or flow. These changes may act upon a diaphragm or a piston which in turn activates the valve. An actuator will stroke the valve depending on its input and set-up, allowing the valve to be positioned accurately, and allowing control over a variety of requirements. Often a sensor element is needed for the position detection in order to allow a precise control of the piston's position. Actuated valves have many uses, starting with the control of the water management in the bath as well as in residential uses such as on/off & pressure control to dish and clothes washers & taps in the home.

Solenoid valves are used in virtually all industries. Especially in the beverage industry, valves with small diameters are used. For example, media-separated flipper valves are used for dosing syrup in a beverage dispenser, or plunger valves for controlling the water and steam in coffee makers. By the length of the triggering

M. Hannig
Bürkert Werke GmbH, Christian Bürkert Straße 13-17, 74653 Ingelfingen, Germany

F. Höhne
Lindner Armaturen GmbH, Kurze Straße 10, 09117 Chemnitz, Germany

S. Langbein (✉)
FG-INNOVATION GmbH, Universitätsstr. 142, 44799 Bochum, Germany
e-mail: langbein@gmx.com

© Springer International Publishing Switzerland 2015
A. Czechowicz, S. Langbein (eds.), *Shape Memory Alloy Valves*,
DOI 10.1007/978-3-319-19081-5_2

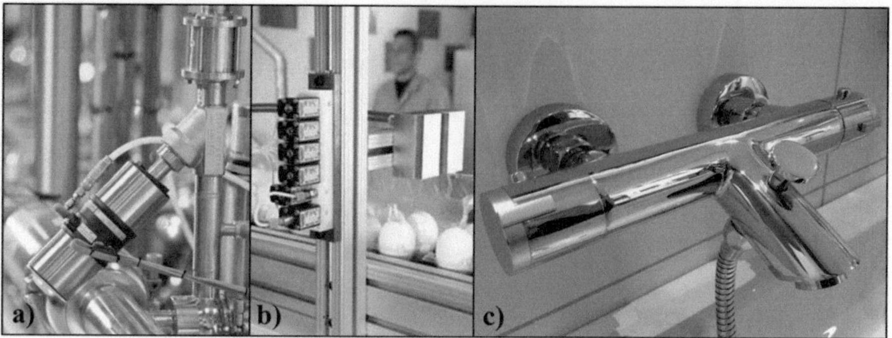

Fig. 2.1 Different applications of valves (**a**) and (**b**) by Bürkert fluidics

signal, the dosing time is often used to control the volume of the dosed medium. Servo-controlled membrane valves with a large diameter and flange connection are used for supplying drinking water in buildings. For dosing, chemical companies prefer direct-acting, media-separated valves made of stainless steel or plastic with medium diameters, while pneumatic applications require miniature plunger valves with a low conveying capacity and diameter. Solenoid valves are ubiquitous—in gas and oil burners, laundries, car washes, and petrol pumps.

Actuated valves are also used in the automotive & aerospace sectors. Air pressure valves are used for example in automotive seats for the enhancement of the comfort. Other active valves can be used in combustion engines and gears for the control of the oil viscosity in reference to the system's temperature.

A thermostatic valve, in contrast to solenoid valves, is a mechanical temperature controller regulating the fluid flow in dependence of the ambient temperature. It is used to keep a constant temperature in systems. Thermostatic valves are used in a variety of products. The best known applications today are thermostatic valves, such as temperature controllers in showers and washbasins or temperature control for heating and air conditioning. The drive elements are today mostly thermal expansion elements. But also thermo-bimetals in some applications are in use. Both drive types convert thermal energy into mechanical energy and are therefore comparable in their function with shape memory actuators.

Figure 2.1 presents some examples of actuated valves in applications like in (a) as proportional fluid control valve for regulation of water flow, (b) micro-valve array for precise dosing application, and (c) a thermal mixing unit for the control of the water temperature in the bath.

2.2   Electrical Valve Systems

2.2.1   *Introduction*

2.2.1.1   SMA: An Actuator Variant with Potential for Industrial Automation

As already explained, the applications for solenoid valves are as diverse as the industries and their media. In addition to solenoid actuators, motor-driven and pneumatic actuators are also available today as alternative valve solutions. Customers and manufacturers are always on the lookout for the technically and economically best solution for their application.

2.2.1.2   A New Approach

The shape memory alloy now presents a new means of actuation. How successful this type of actuation can be depends on

- The manufacturing costs of the SMA actuator
- The ability to integrate it in the industrial requirements (voltage supply, environmental conditions, certifications)
- And the additional benefit for manufacturers and plant operators

2.2.1.3   Costs of the SMA Actuator

Considering the actuator alone, a shape memory alloy actuator is the most economical actuator type. The high-alloy material features excellent anti-corrosion properties and is available today in consistently high quality [1]. Shape memory alloys are available from different manufacturers as inexpensive wire and rod material. Compared with a simple wire or rod actuator, a solenoid or motor actuator with its magnetic components consisting of the armature, stator, baffle plates and coil is very complex and expensive. In a brief analysis therefore the SMA actuator offers the best cost advantage.

2.2.1.4   SMA and Industrial Requirements

If one considers not only the SMA actuator itself, but also the basic requirements of industry, it becomes apparent that the SMA actuator involves requirements and limitations that restrict its use with respect to both technical and economic aspects.

2.2.1.5 Voltage Supply

The typical voltage supplies used in process manufacturing and in machine design are 12…24 V DC/AC and 110…240 V AC. Normally, low electrical currents are used to minimize the costs of supply lines and junction boxes. The requirements for an SMA actuator are just the other way around; it requires low voltage, e.g. 3 V, and a relatively high current, e.g. 0.5…2.5 A for prompt actuation. Current converters present a possible solution, although they result in higher manufacturing costs for the actuator.

2.2.1.6 Protection Type (NEMA, IP)

A high level of protection against dust and moisture is a standard requirement in all industrial areas today. Protection types IP54 and IP65 (NEMA 4X) are considered standard. To a lesser degree, but just as important, are the protection types IP67/68 and IP69K for especially rough environments and pressurized water. Solenoid and motor-driven actuators meet such requirements through hermetic sealing of the electrical and, if necessary, of the mechanical parts. Limitations in performance with respect to the switching or operating behavior do not occur—due to the known sealing technologies. In an SMA actuator the required protection types must absolutely dissipate the heat from operation, since this has a decisive effect on the switching time. The known and time-tested methods for achieving the protection type in solenoid and motor-driven actuators through encapsulation or cavity ventilation with PTFE filters are only of limited use here. Since the current-carrying wire or rod itself is the actuator, it must be encapsulated both electrically and mechanically, yet still must be able to dissipate sufficient heat. This results in increased costs for materials and design and requires new concepts for achieving the protection type.

2.2.1.7 Ambient and Operating Temperatures

The operating temperature range for solenoid and motor-driven actuators is specified in the standard as −40 to +75 °C (−40 to +160 °F). In special versions, extreme operating temperatures of −60 °C (°F) or +250 °C (480 °F) are also possible. Ambient and media temperatures can easily cause the temperatures in and around the body to reach the maximum temperature. Depending on the system and design, temperatures can also temporarily exceed the specified continuous operating temperature. Solenoid and motor-driven actuators can withstand these temperatures with minimal loss of performance. The effects of ambient temperature present a special challenge for the switching speed and repeat accuracy. An SMA actuator without electronic temperature compensation is hardly conceivable here.

2.2.1.8   Certifications

There are no test and evaluation criteria for most standard certifications in industry, since no products have been submitted to the corresponding organizations and committees for testing. Evaluations and tests in accordance with SIL, the machine directive, explosion protection, gas or chemical applications therefore have not been created. Pioneering work by the individual manufacturers is necessary here.

2.2.1.9   Customer Benefits and Summary

The low costs of the SMA actuator will essentially be cancelled out by the costs of current conversion, electronic temperature regulation, and protection type. The manufacturing costs of SMA actuators for industrial automation would therefore be approximately equivalent to the costs of solenoid or motor-driven actuators. Despite comparable manufacturing costs, however, SMA actuators offer additional benefits for the customer:

- The actuator causes no EMC problems in system operation
- Alternative low-cost implementation of a solenoid control actuator
- Safety valve character in case of power outage (spring reset)
- Preventive functional monitoring of the switching function (SIL, machine directive)
- Smooth valve operation due to minimal pressure and closing impacts
- Low weight and volume

2.2.1.10   Summary

Despite the technological competition there are promising opportunities and niches for the SMA actuator. Especially, the increasing industrial requirements with respect to asset management, functionality, self-monitoring, and closing safety offer points of departure for new developments. Certification barriers, protection types, or loss of performance due to ambient temperatures can be solved if the specifications are correct. It is therefore advisable for manufacturers and customers to first identify solutions in the standard applications of industrial and building automation as a startup product.

2.2.2   Solenoid Valves—Basics

2.2.2.1   Physical Basics of Solenoid Drives

In the following chapter, the state of the art of electric active valves is discussed exemplary referring to solenoid-driven valves. Solenoid valves use the electromagnetic effect, which is based on the magnetic induction which is discussed here in a

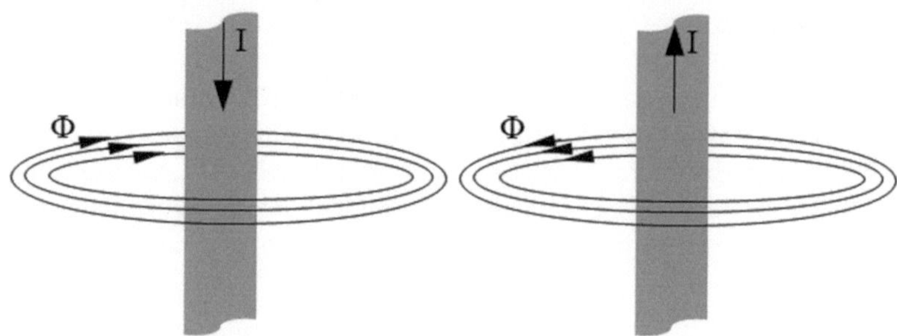

Fig. 2.2 The basics of the electromagnetic field

brief version. The magnetic fields can be inducted in magnetizable metals such as Iron, Nickel or Cobalt.

While an electric current flows in a conductor, an electromagnetic magnetic field is created due to Ampere's law. Strong electromagnetic fields can be produced if a magnetic core of a soft ferromagnetic material, such as iron, is placed inside the coil. A core can increase the magnetic field to thousands of times the strength of the field of the coil alone, due to the high magnetic permeability μ of the material. Figure 2.2 shows the reversibility of the magnetization direction by changing the direction of the current flow (I).

The integral of the magnetizing field H around the closed loop of the field is equal to the sum of the current flowing through the loop.

If the conductor, i.e. a copper wire, is winded around a ferromagnetic core, the magnetic field is proportional to the turns in winding (N) and the current in the wire (I).

The magnetic flux density (B_L) is vital for the calculation of the magnetization of the gap between the core and a moveable piston (core part), as displayed in Fig. 2.3.

For an electromagnet with a single magnetic circuit, of which length L_{core} of the magnetic field path is in the core material and length L_{gap} is in air gaps, Ampere's law reduces to Eq. 2.1:

$$F_M = \frac{B_L^2 \cdot A_L}{2 \cdot \mu_0 \mu_r} \tag{2.1}$$

With: F_M = pulling force of an electromagnet

μ_0 = permeability of free space = $4\pi 10^{-7}$ Vs/Am

μ_r = permeability factor = 1 for air, <1 for field-weakening materials (e.g. Copper), >1 for field reinforcing materials (e.g. Palladium) and >>1 for ferromagnetic materials (e.g. NdFeB).

A_L = cross section area of each air gap

B_L = Magnetic flux density = Φ/A_L

Δs = air gap length

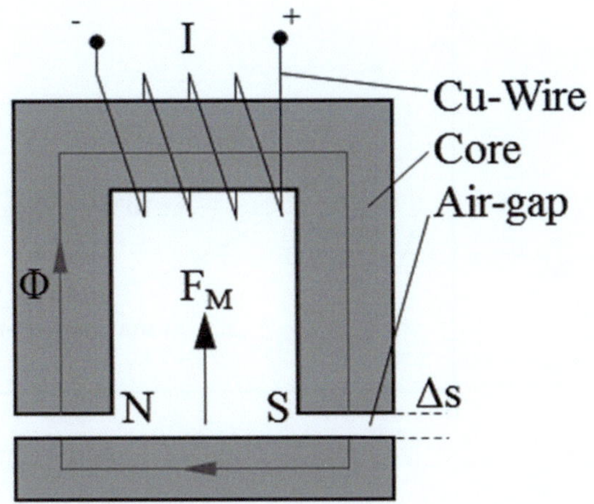

Fig. 2.3 Schematic of an electromagnet

The magnetic flux density is a factor of magnetizing field (H). With respect to Eq. 2.1 the major factors for the pulling forces are the constructive dimensions of the electromagnet and the air gap length due to Eq. 2.2:

$$\Theta = N \cdot I = H \cdot \ s \cdot k \tag{2.2}$$

With: $\Theta = N \cdot I$; $N =$ turns in winding, $I =$ current
  $H = B/(\mu_0 \mu_r)$
  $k =$ number of air gaps

If Δs increases, the necessary current level rises in order to maintain the calculated magnetic force in Eq. 2.1. As a result, the magnetic force decreases at constant current level and with rising air gap length. Figure 2.4 shows an exemplary measurement of force level over the air gap (stroke position) of a solenoid drive. As discussed, the force level changes significantly with the piston's position in many solenoid drives. However, there are also high-performance solenoid actuators available which enable nearly constant force levels over the stroke but at high constructive effort and increased costs. Due to thermal effects, the recommended duty on times are related in percentage to a maximal on time. With the increase of the maximal on time, the usable force decreases over the stroke (Fig. 2.5).

2.2.2.2 Solenoid Valves: Industrial Demands and Standard Design Types

Solenoid valves are and will remain an indispensable component for controlling liquids and gases. No industry can get by without using them. Solenoid valves can be found in aerospace and automotive technology, in household appliances, in pharmaceutical and food technology and in power and sewage treatment plants [2]. The simple and time-proven actuator concept of the solenoid coil and armature is successfully used today billionfold.

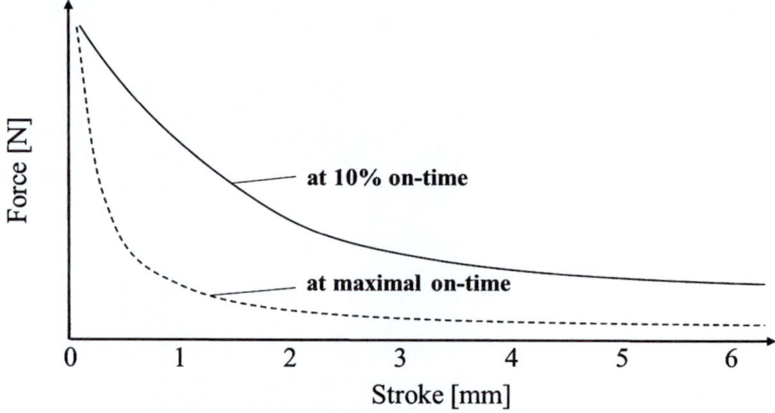

Fig. 2.4 Exemplary force decrease characteristic with increasing stroke of a solenoid drive

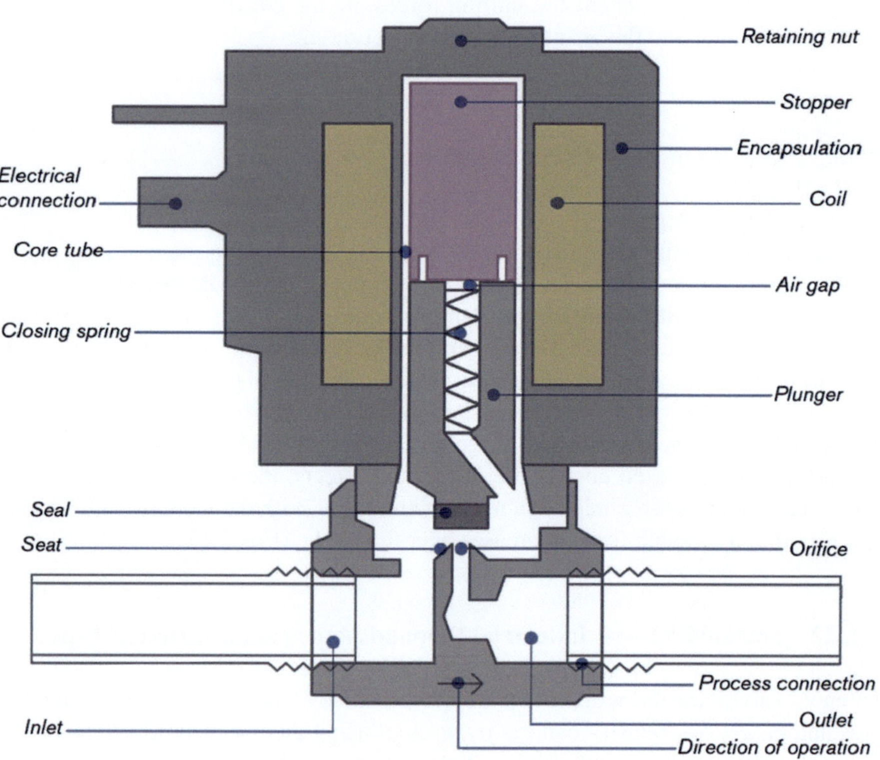

Fig. 2.5 Principle of a direct actuating solenoid valve

The greatest benefit to both manufacturers and customers is the simple, cost-effective and reliable design of the actuator, and especially the contactless transmission of force. Solenoid valves with spring reset are considered by experts to be "safe technology" for switching to a safe operating state in the event of a power outage, for example. Extreme and harsh ambient conditions with temperatures from $-100\,°C$ to $+250\,°C$, explosive atmospheres, salt spray or IP protection class IP 69 K are standards for solenoid coils today. The low sensitivity to electromagnetic and radioactive emissions expands the area of use to applications with maximum safety requirements. The current performance spectrum of the coil actuator includes simple On/Off, solenoid control and safety valves.

Due to its worldwide use in virtually all industries the coil actuator is subject to numerous international and national standards. The corresponding certifications and certificates are granted on the basis of well-known standards. Depending on the standard, only the electric actuator or the entire valve is included in the test. Typical examples for this are certifications such as ATEX, IEC Ex, FM, NEPSI, UL, CSA, GOST, VDE, and SIL. The solenoid valve user of today can be assured of finding suitable valves on the market with his required certification and requirements.

The coil requirements focus today following major factors:

- Supply voltage: often in the range between 12 and 24 V, some up to 230 V
- Control type: constant level controller, two-point controller, pulse wide modulation controller
- Protection type: protection against environmental factors
- Electrical connection: standardized installation and service possibilities
- Certifications

2.2.2.3   The Main Requirements of Solenoid Valves

The advantage and benefit of a valve is often not a question of the actuator type. The actuator is a means to an end and therefore secondary. What the user wants above all is the reliable opening, closing, locking, distributing, dosing or release of a line or container. To achieve this, the valve must fulfill many customer requirements.

Fluidic requirements:

- Medium (liquid, gaseous, aggressive, contaminated, abrasive, steam):
  The medium is the fluid which is transported and controlled by the valve. Often it is the valve user's product (e.g. water) or a medium for actuating the user's machines (e.g. oil for hydraulic actuators).
- Media temperature:
  The medium's temperature has to be regarded during the utilization of a valve. Due to thermal transport effects, the thermal energy can be conducted even to the actuator itself. If the actuator is sensitive to thermal fields, its actuation can be affected.

- System pressures:

  The system pressures can occur as switching pressures (e.g. for pilot valves) and media pressures. Especially considering direct working valves, the pressure affects directly the actuator and is one of the major design parameter of the drive element.

- Diameter:

  The diameter limits the flow rate and the pressure levels in a valve.

- Binary or solenoid control valve:

  The control type is essential for certain applications. While a binary (on/off) valve only focuses the two final states of the valve (opened and closed), the controlled valve has the ability to regulate either the pressure or the flow rate of a valve by regulating the valve piston's stroke.

- Switch-on time (related to system dynamics):

  Is the time which is set by the user for opening the valve.

- Switch-off time (related to system dynamics):

  Is the de-activation time which is needed by the valve to shut-off.

- Max. switching frequency:

  Is the frequency resulting from a continuously on/off switching.

2.2.2.4 Subordinate Requirements of Solenoid Valves

Corresponding to the user's needs, some technically subordinate factors are also interesting for the application of valves in industrial processes and products:

- Pressure differential between valve inlet and outlet
- Flow rate in Kv or QN
- Materials used for the design (related to the corrosion behavior)
- Process connection
- Media separation
- Certifications

In the end, it is always the fluid requirements and the required certifications that determine the power, size, and power consumption of the electromagnetic actuator. Due to the large number of applications and corresponding adaptations there are countless customized solutions. All of the solutions, however, can be divided into two standard types based on their operating principle and function.

The main differentiation of valve design focuses the usage of the actuator in relation to the switching piston. There are direct-acting valves, which generate the moving force directly by a solenoid or pilot-controlled valves which is supplied primarily by the differential pressure in the valve. Figure 2.5 shows the principle of a direct working valve according to the state of the art.

The solenoid drive is used in a direct actuating valve for the movement of the plunger (piston). If the coil is triggered by electrical current, the stopper is magnetized (core) and the plunger is moved. As result of this movement, the seal element in the valve seat is displaced, opening the fluid flow from the inlet to the outlet of

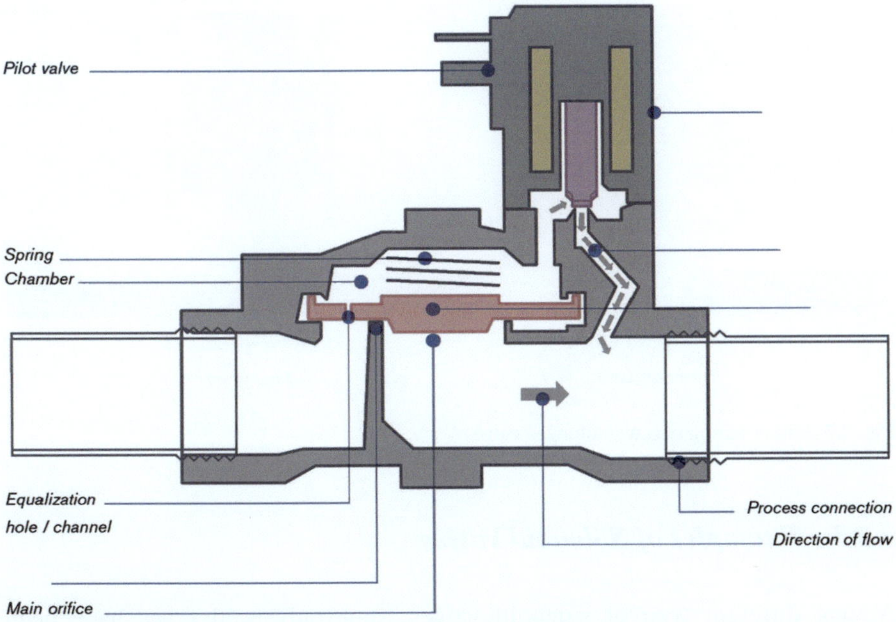

Fig. 2.6 Principle of a valve which is actuated by a pilot solenoid valve

the valve. The orifice limits the pressure and flow volume of the valve, and is one of the major factors concerning the solenoid actuating forces. After switching off the coil, the closing spring moves the plunger back toward the seat, closing it with the seal. In order to enhance the performance of the valve, the movement of the plunger is supported by the inlet fluid pressure. Hence, the solenoid drive core is under the inlet fluid pressure. This constructive type is not a medium-separated direct valve.

If the volume and/or pressure of the fluid increases, the necessary switching forces which have to be generated by the solenoid will rise disproportionately to the volume and pressure increase. In order to control such high force regulations of the fluid, valves with a smaller pilot valve are used. Figure 2.6 presents a pilot valve that consists of two valve elements.

Opening large orifices using the direct acting method would require enormous and expensive coils. Therefore, servo-assisted valves use the power of the fluid to open the flow channel by controlling a small pilot channel to alter the forces on a larger main seal. The seal has a small equalization hole in which conducts the inlet pressure parallel to the main orifice to the pilot valve. The core of the pilot valve has effecting areas. The pressure affects on these areas as a resulting force against an inserted closing spring. If the coil is triggered, the resulting force and the solenoid pulling force are stronger than the closing spring of the pilot valve. If the pressure arrives in the right part of the main valve, the balance of the main spring at the main seal and the pressures is shifted. As a result, the main seal opens (either by displacement or by deformation in some designs).

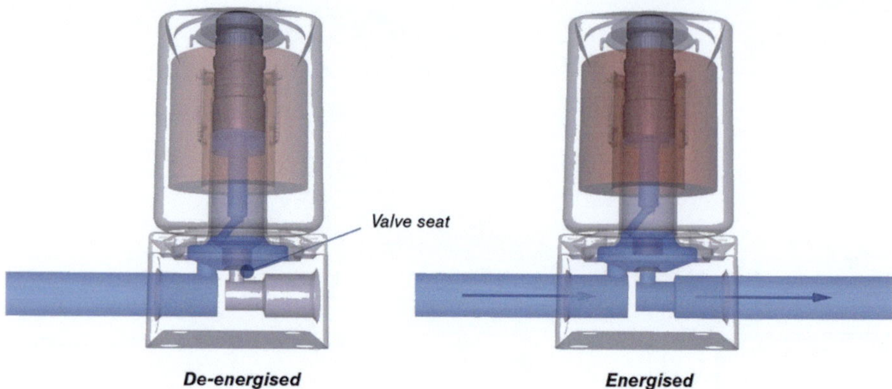

Fig. 2.7 Direct-acting two-way plunger valve

2.2.3   *Examples of Solenoid Valves*

Among different type of solenoid valves, some advanced types have been released on the market which enable special features and are suited for certain applications:

2.2.3.1   Direct-Acting 2-Way Plunger Valve

Function: The main components of this valve type (see Fig. 2.7) are a coil, a closing spring, a valve body cover, and the valve body with the seat. Without current the path to the outlet is blocked (normally closed), since the closing spring, supported by the pressure of the medium, presses the plunger onto the valve seat. If current flows through the coil, the latter generates a starting force, which pulls the plunger and the seal against the spring force and draws the medium upward. The channel is opened for the medium.

Application: This cost-effective valve type is used in universal applications for neutral and clean liquids, gases, and vapors. Versions with special high-quality materials also allow their use in mildly acidic and alkaline solutions. The direct-acting 2-way plunger valves therefore can be used for diverse applications, such as shut-off, dosing, filling, and ventilation.

Special features: Due to a spring-damped seat seal, these valve types have a long service life. Especially noteworthy are the increased switching cycles and service life due to sliding ring bearings. These products are also suitable for high pressure and temperature range.

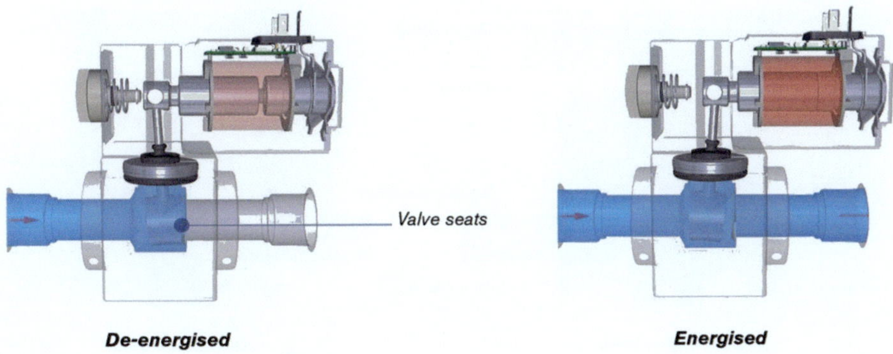

Fig. 2.8 Direct-acting toggle valve

2.2.3.2 Direct-Acting Toggle Valve

Function: The valve operates according to the lever principle and can therefore also directly switch large diameters. It is available both as a 2/2-way and a 3/2-way valve.

The armature acts horizontally on a fixed coupled toggle arm. The sealing cylinder located on the lower lever is pressed by the horizontal motion onto the valve seats. The plastic encased metal lever comprises one unit with the gas-tight lead-through. Due to this design the actuator is media-separated from the fluid body. The valve system is shown in Fig. 2.8.

Application: Media separation makes this valve especially suitable for use in critical acidic and alkaline solutions or in media that contain particles. Due to the large diameters it is often used as an emptying and mixing valve.

Special features: The energy-saving version with power reduction uses the double coil technology with integrated cast electronics. They are certified worldwide as AC, DC and UC versions and fulfill the voltage requirements for European rail transport.

These valves are equipped with a locking service-friendly manual override and offer the capability of potential-free electrical feedback of the switching position.

2.2.3.3 Direct-Acting Pivoted Armature Valve

Function: This type of valve (see Fig. 2.9) uses a pivoted armature, a flexible separating diaphragm, two valve seats and one coil. They can be used both as 3/2-way and 2/2-way versions. Under voltage the pivoted armature is pulled against the force of the spring and the path between P and A (outlet) is opened. At the same time the

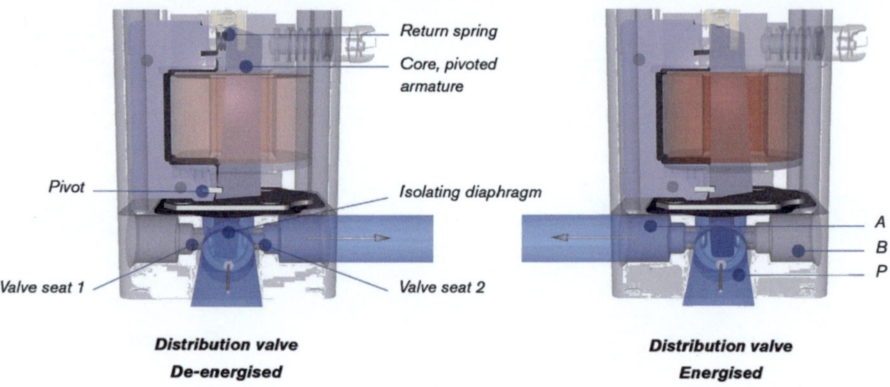

Fig. 2.9 Direct-acting pivoted armature valve

channel between P and B is closed. Without current the pivoted armature closes valve seat 1 and the medium can flow between connection P and B.

Application: The use of a separating diaphragm, which separates the media chamber from the magnetic system, makes it possible to use these valves for the control of corrosive, contaminated and aggressive fluids as well as for vacuum. These valves are equipped with a lockable manual override and offer the unique capability of electrical feedback of the switching position, which results in increased safety.

Special features: Versions for use in explosive areas are available, as well as different materials for media-contacting components. Decades of engineering experience make it a highly reliable, low-maintenance valve.

2.2.3.4 Direct-Acting Pivoted Rocker Valve

Function: The special feature of direct-acting rocker valves is that all orifices are located on one plane in the valve body. One type of a rocker valve is shown in Fig. 2.10. This functioning principle is suitable both for 2/2-way and for 3/2-way valves. The valve consists essentially of four main elements: magnetic coil, spring, rocker, and two valve seats. In a 2/2-way valve a spring presses one side of the rocker against one of the two valve seats, which is sealed in this way. As soon as the coil is energized, supported by an additional spring it pulls the part of the rocker that was previously pressed against the valve seat. This creates a "rocking motion" and the valve seat opens. At the same time the other side of the rocker is pressed against the blind seat. Since it is not completely closed, medium can now flow unhindered through the valve.

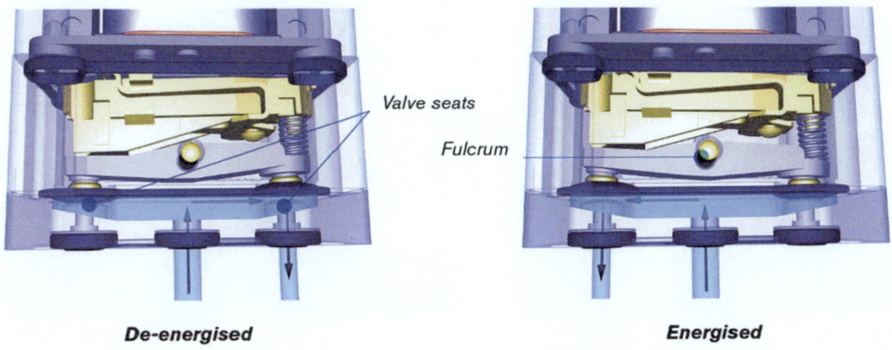

Fig. 2.10 Direct-acting pivoted rocker valve

Application: In the standard version such valves are used for the control of neutral gases and fluids. They can be used either as single systems or as pilots for pneumatic valves in normal or explosive environments. Rocker valves can also be equipped with a separating diaphragm to separate the valve mechanism from the medium. This makes them suitable also for switching aggressive and high purity media.

Special features: The design with an integrated membrane is the basis of the MicroFluidics series, in which minimal internal volumes and excellent flushability are especially important.

2.2.3.5 Diaphragm Valve with Plunger Pilot Control

Function: This functioning principle (see Fig. 2.11) uses a direct-acting plunger valve as the pilot valve and a flexible diaphragm as the seal for the main seat. As soon as the pilot valve opens, the fluid chamber above the diaphragm is emptied. The media pressure within the diaphragm raises the diaphragm and opens the valve so that the medium can flow. If the pilot valve is closed, the media pressure above the diaphragm builds up again through the small compensation opening and the closing process is supported additionally by the compression spring. A minimal differential pressure between the inlet and outlet is necessary for complete opening and closing.

Application: The main areas of application for this pilot-controlled solenoid diaphragm valve are clean liquid or gaseous media such as compressed air, water, hydraulic oils, etc. The small pilot valve makes them less expensive than direct-acting valves for use with higher pressures and larger diameters.

Special features: Since pilot-controlled valves (also: servo-assisted valves) have only a small pressure equalization hole in the diaphragm, they are susceptible to dirt particles and crystallizing media, which can clog the hole. Bürkert valves are

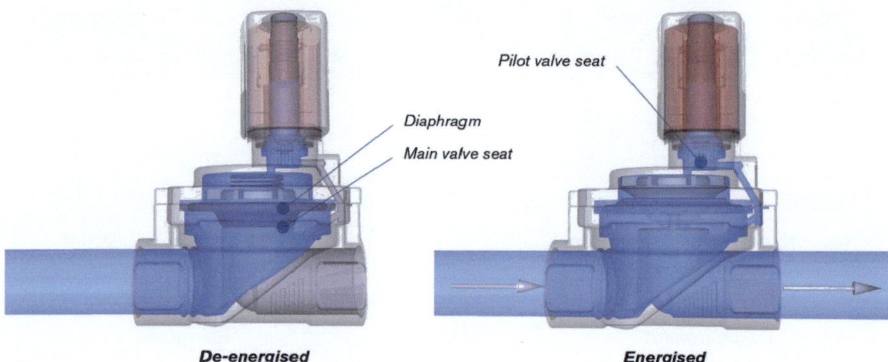

Fig. 2.11 Diaphragm valve with plunger pilot control

designed as soft stop valves. Ex or low power versions are virtually unproblematic compared with direct-acting valves, since the pilot control valve only has lower power consumption.

2.3 Thermostatic Valve Systems

2.3.1 Introduction

Thermostatic valves are now dominated by expansion elements like wax elements. These drives are inexpensive manufactured in very large quantities. This fact makes it difficult for shape memory alloys expand into these markets. Although thermal shape memory actuators are mostly designed as coil springs and so are very easy to integrate, the target cost can often not be achieved. For this reason, the shape memory applications are found only in niche products; or the technical specifications as the installation space promote the use of shape memory drives. In the field of thermostatic valves shape memory alloys are therefore able to enforce only through a price reduction.

However, the increasing monitoring of technical systems does not stop at thermostatic valves. In this case, expansion elements will lose the connection, since electrical control of these systems is very complex. The expansion elements must be heated by an externally attached heating element and be monitored by an external sensor. With underfloor heating systems, the valve drive is realized in this way. The possibilities offered by shape memory actuators in this context provide a much greater relevance of these drives. In summary it can be stated that the development of thermostatic valves to intelligent systems requires the use of new drives. Shape memory actuators are available for this purpose.

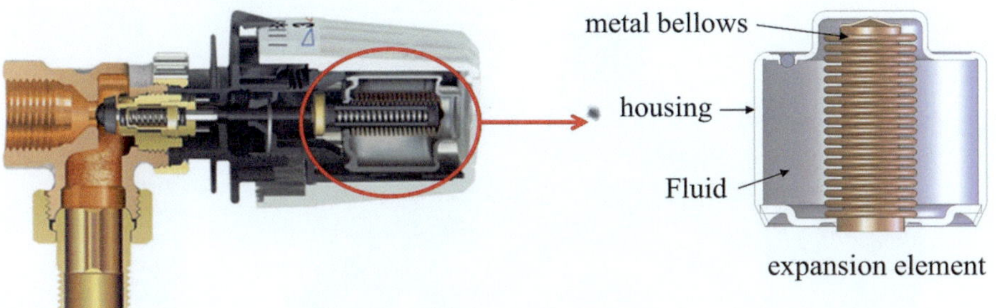

Fig. 2.12 Thermal heating thermostat and expansion element [1]

2.3.2 *Examples of Thermal Valves*

2.3.2.1 Heating Thermostat Valve

A heating thermostat valve (see Fig. 2.12) is an individual room temperature control unit, which has a temperature sensor or a thermal actuator in the form of an expansion element inside. The Expansion element is detailed in Fig. 2.12 on the right sight. The expansion element expands according to the room air temperature. This change in length is transferred to a transducer piston and via this to the valve, which thus changes the flow cross-section of the heating forward travel and the flow rate. A return spring is used to open the valve during the cooling of the expansion element. Continuous control of the heating fluid provides a constant room temperature. Without the mounted thermal expansion element, the heating valve is fully opened.

The air temperature is selected by turning the adjustment head. Turning the head to the right, the whole head is moved with expansion element and transmission piston closer to the valve body. The passage of the valve is reduced by the heated expansion element. The heat output of the heating is reduced and the air temperature decreases. The setting at a medium level (usually "3") means a room temperature of about 293 K. At falling room temperature, the valve is opened again. Since thermostatic valves are proportional controllers, a small deviation from the target value remains and so the actual value oscillates within a range of about 2 K.

For the control of underfloor heating return temperature limiter is often used. This thermostat valve does not measure the room temperature, but the water temperature in the return system of the underfloor heating.

Because of the variety of applications and a good cost-usage ratio thermostat elements have become an important part of the building installation. The dynamics of control behavior are nearly irrelevant due to the inertia of the whole system.

Fig. 2.13 Thermostatic fluid controller based on SMA

2.3.2.2 Thermostatic Mixing Valves

Thermostatic mixing valves as used in showers, bathtubs or basin mixers allow a constant outlet temperature of the mixed water, regardless of the temperature and largely independent of the pressure of the water supply lines. The hot water temperature can be kept constant with a variation of 1–3 K from the set temperature. For safety reasons the thermostats have a lock on the adjusting knob which is set to 311 K. There are also special versions in which the mixing chamber is outside cooled by inflowing water so that the valve's outer shell remains cool.

Inside the mixing thermostats have a controlling element which is surrounded by the mixed water. The controller consists either of bimetallic wire which has been wound into a helical spring or a thermal expansion element.

Shape memory alloys are already used in thermostatic mixing (see Fig. 2.13) since several years already. In particular, the fast response and small installation space of SMA drives are beneficial here. A more detailed description of these systems is presented in Chapter 11.1.

2.3.2.3 Scald Protector

For the implementation of scald protection valve a very fast regulation is needed. For the immediate closure of the hot water supply (in case of failure of the cold water flow) a fast thermal actuator at moderate forces is required. Such a valve is shown in Fig. 2.14.

Due to the security relevance in scalding in relevant standards such as EN1111 the properties which must be satisfied through the thermostatic valve and the number of appropriate test cycles are defined. For example, the valve must pass 70,000 cycles

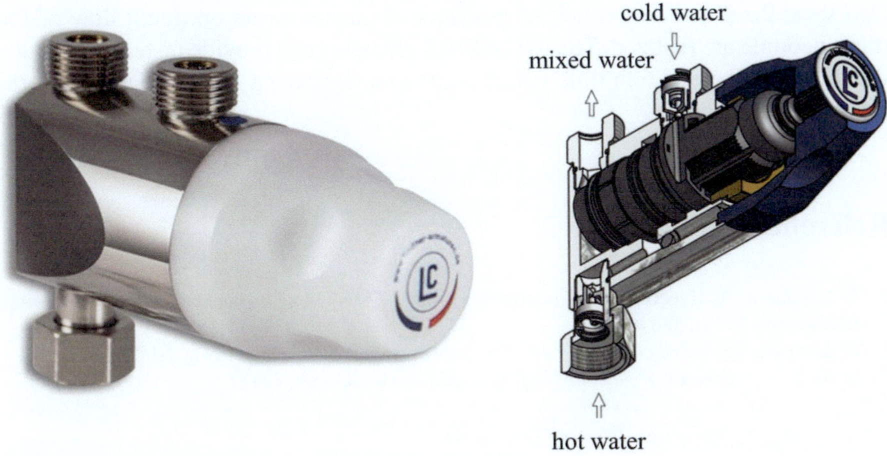

Fig. 2.14 Mini thermostats for scald protection by Lindner Armaturen GmbH

per EN1111. These tests are important because the thermostatic safety, especially in the public sector, must ensure that even most vulnerable users, such as in kindergartens, in hospitals or in nursing homes, will be protected.

As the thermal expansion elements are available in miniaturized design with progressive development now, their use is popular in cartridges with ceramic discs. A general protection in all plumbing, heating and pneumatic applications is technically possible.

A distinction is generally made between 2 modes of operation: permanently set means that the valve body is positioned with expansion element in the supply of hot and cold water and that expansion increases with the mixing temperature. The valve is opened and closed, that a fixed temperature is maintained. For freely selectable temperatures the position of the valve block is modified with thermal expansion elements so that the maximum mixed water temperature can be adjusted to warmer or colder levels.

Usually the cold water inlet and thus scald protection cannot be blocked by the user. In today's applications in which thermal disinfection of line equipment is needed, first thermostatic valves are already being offered with an anti-scald deactivation possibility. In that application case, the hot water flow can be released through the valve. This is realized via bypass functions or settings using special tools. Mini thermostats allow both: to the secure electronic faucets with only one mixed water outlet and the thermostatic safety of the hot water supply.

By structural design of the water flow to or around the thermal expansion element its reaction time is affected. In case of failure of the cold water inlet, mini thermostats must close within one second in order to ensure effective protection against scalding. In addition to the adapted geometry of the expansion element, also specialized solutions such as guidance elements for an optimum flow of the thermocouple are realized. The use of SMA elements can provide faster heat uptake in relation to the reaction time which leads to a significant improvement concerning reaction dynamics.

References

1. S. Langbein, A. Czechowicz, *Konstruktionspraxis Formgedächtnistechnik* (Springer Vieweg, Mannheim, 2013). 3834819573
2. R. Merrick, *Valve Selection and Specification* (Springer, New York, 1991). 0442318707
3. D. Will, N. Gebhardt, *Hydraulik* (Springer, Berlin Heidelberg, 2011)

Chapter 3
Introduction to Shape Memory Alloy Technology

Alexander Czechowicz and Sven Langbein

3.1 Basics of the Shape Memory Effect

Alloys with the ability to revert to their programmed shape through thermal activation are known as shape memory alloys (SMA). The diffusion-less phase transformation of the low-temperature phase (martensite) to the high-temperature phase (austenite) is the basic characteristic of shape memory alloys. During this martensitic transformation, neither the chemical composition of the alloy nor the collocation of atoms changes. The conversion from the austenite phase to the martensitic phase starts below martensite start temperature (M_S). By way of further sub-cooling, more martensite crystals form in the structure. From an automation perspective, the phase transformation effect is activated by channeling thermal energy into the system. The phase transformation is reversible and not time-dependent, as it only distorts the atomic lattice without causing long-range migration of atoms [1]. Functionally speaking, the phase transformation is an application of stress to both the material and the temperature, leaving the conversation rate unaffected by the duration of the two acting factors. The stress needed to induce phase transformation is dependent on the M_S temperatures and on the ambient temperature. Therefore, the M_S temperature increases in equal measure with the applied mechanical stress. In contrast to hardening processes in steels, only view-irreversible defects may be generated in SMA so that the reversibility of the conversion for the maximum number of cycles is maintained [2]. Figure 3.1 illustrates the phase

A. Czechowicz (✉)
Zentrum für Angewandte Formgedächtnistechnik, Forschungsgemeinschaft Werkzeuge und Werkstoffe e.V., Papenberger Straße 49, 42859 Remscheid, Germany
e-mail: czechowicz@fgw.de

S. Langbein
FG-INNOVATION GmbH, Universitätsstr. 142, 44799 Bochum, Germany
e-mail: langbein@gmx.com

© Springer International Publishing Switzerland 2015
A. Czechowicz, S. Langbein (eds.), *Shape Memory Alloy Valves*,
DOI 10.1007/978-3-319-19081-5_3

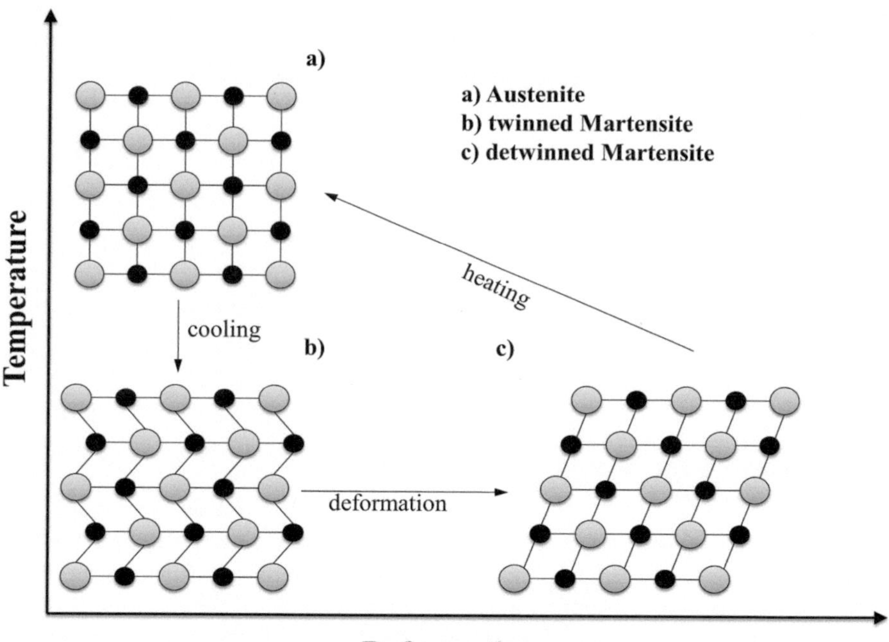

Fig. 3.1 The martensitic transformation in shape memory alloys

transformation schematically. By cooling from austenite (a) to martensite, the twinned martensite is formed (b), and, through displacement of the interfaces, undergoes deformation to (c).

The twinned condition occurs when within the martensite grid several martensite groups arise with different orientations. Due to the material deformation, these martensite groups align in the preferred direction. This so-called detwinning procedure assumes highly mobile martensite interfaces, set by the acting mechanical stress that stretches the material. The atomic grid transforms to the austenitic structure when heated. The phase transformation does not occur at a specific temperature but within a temperature interval. Figure 3.2 shows the phase transformation degree (ξ_A) as a material temperature function.

The transformation process can be described by four characteristic temperatures (also called phase transition temperatures or PTT), which can be determined from diagrams using a tangent method. With a constant mechanical load, the martensitic transformation begins at the so-called austenite start temperature (A_S). Monoclinic martensite is converted to body-centered cubic austenite. The transformation is complete when it reaches the austenite finish temperature (A_F) and the material is fully austenitic. The reverse transformation starts at the martensite start temperature, which is significantly lower in binary nickel-titanium shape memory alloys than the AF temperature, and ends at the martensite finish temperature, which is

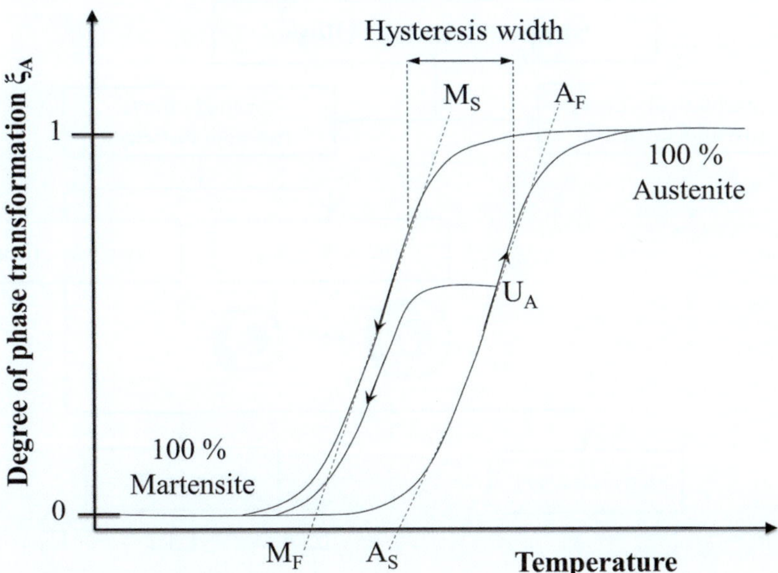

Fig. 3.2 The martensitic transformation in shape memory alloys over temperature

also significantly lower than the As temperature. Thus, the phase transformation behavior shows a thermal hysteresis. The hysteresis is a result of internal losses, such as friction and spontaneous relaxation at the atomic level during the structural transformation [3]. Hysteresis properties, such as shape and width, depend on multiple factors, for example alloy composition, thermo-mechanical pre-treatments and mechanical load. During cooling, the structure transforms gradually from individual areas, and returns to the formation of differently oriented martensite groups (twinned structure). In cases where the phase transformation is not complete, the reverse transformation begins in a mixed phase, consisting of martensitic and austenitic phases. The reverse transformation then unfolds as more of a curved branch, starting at the termination point (Ua) instead.

3.2 Shape Memory Effects

An overview of utilizable SMA effects is shown in Fig. 3.3, including examples of application. While pseudo-elastic SMA are well-known in medical applications like stents and highly flexible consumer products like clothing, thermal SMA are increasingly being used in thermal clamps and electrical actuator applications as well. A more detailed explanation of individual effects is presented throughout the following sections.

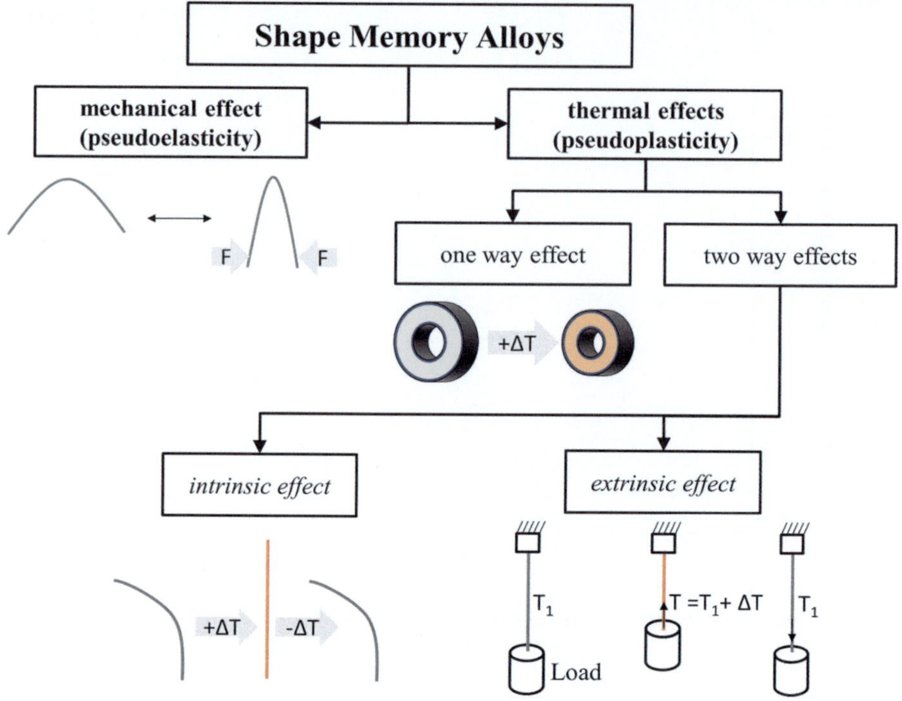

Fig. 3.3 Overview of shape memory alloy effects

3.2.1 The One-Way Effect

SMA elements using the one-way effect can be deformed. After subsequent heating and without applying any extrinsic mechanical load, the element transforms back into its previous shape. Considering the deformation as a function of mechanical stress and temperature, as shown in Fig. 3.4, helps to understand the effect better. The application of stress results in a pseudo-plastic deformation. This deformation appears to be plastic, while the SMA element is relieved. This is called "pseudo" plastic, because it can be reset by shape memory transformation, which is caused by thermal activation. When heated, the austenite start temperature (A_S) is reached. Upon further heating, up to the austenite finish temperature A_F, the material transforms into the austenitic state completely. All atoms adopt the structure of the body-centered cubic grid, which undoes their deformation. Now, cooling occurs down to the initial temperature, at which point the structure transforms back into twinned martensite. During the cooling process, however, the macroscopic structure of the element is not deformed again. This effect causes the material to "remember" exclusively its high temperature form. The one-way effect is suited for applications that generate high forces, or for on-off actuating movements in the lifecycle of

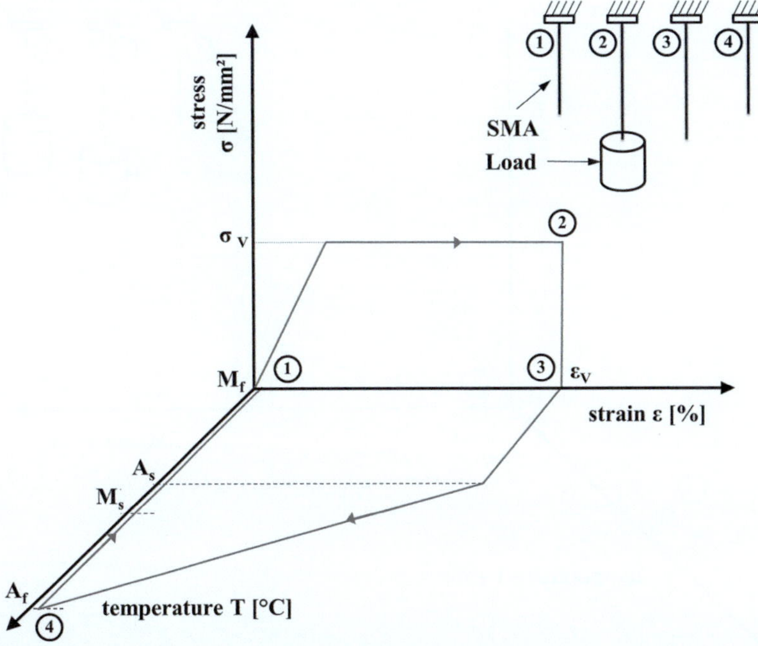

Fig. 3.4 Characteristic of the SMA one-way effect

technical systems. Industrial applications mainly stretch across the field of connection technology (e.g. shrink sleeves) and in space (e.g. one-time opening of flaps or solar sails in satellites).

3.2.2 The Extrinsic Two-Way Effect

The extrinsic two-way effect describes shape memory elements that are continuously under mechanical preload. Figure 2.3 shows the extrinsic two-way principle exemplified by a shape memory wire. When the wire is heated, it contracts and lifts the attached weight. Figure 3.5 additionally illustrates the schematic stress–strain–temperature diagram of the effect with marked reference points that connect back to the main diagram. The unoriented martensite (1) gets deformed by external stress until it reaches an apparent yield point (2). At this point, the martensite is detwinned. This pseudo-plastic deformation, which remains intact even after the SMA element is released, is called pseudo-plastic elongation. When the SMA element is heated, the phase transformation starts at A_S temperature and the martensite structure is transformed into austenite (3). The nature of this transformation can be attributed to the orientation relationship between the austenite and martensite grid. Because of the continuously applied mechanical stress, for example by an external load, the material is stretched directly into the detwinned martensitic state during the cooling process

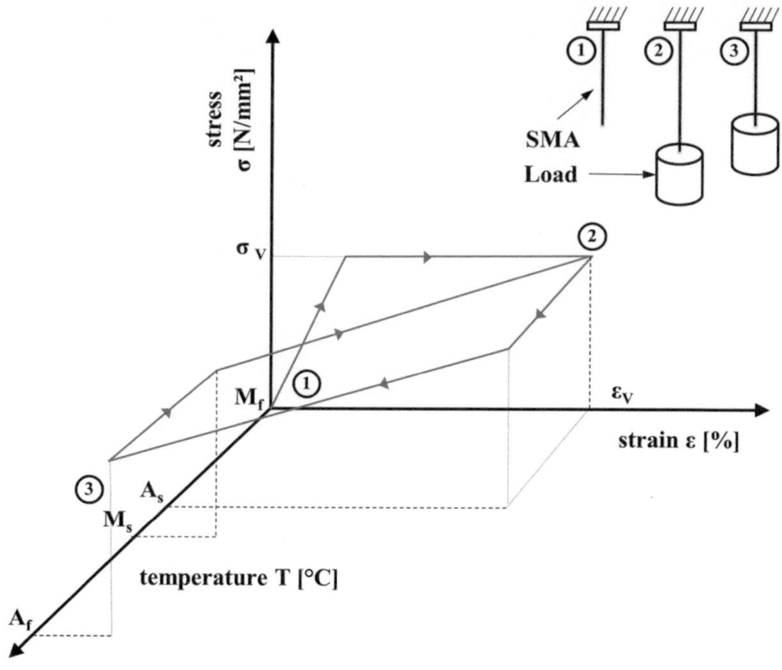

Fig. 3.5 Characteristic of the SMA extrinsic two-way effect

(2). This effect is suitable for cyclic operations in valve drives. The phase transformation from martensitic to austenitic state can be triggered via Joule heat generation in order to contract the SMA wire by electrical current, as presented in the main diagram. The shape memory alloy exerts a mechanical reaction to the applied stress level. The product is a resulting force, or a resulting stroke, which can be used for mechanical work. In the actuator application, SMA can normally be switched between states (2) and (3). It is also possible to target intermediate positions along the hysteresis curve and maintain those by means of control strategies. This allows for the use of SMA actuators as proportional drives.

3.2.3 The Intrinsic Two-Way Effect

The intrinsic two-way effect describes shape memory alloys that show reversible transformation behavior both during heating and cooling. This behavior is introduced into the material by thermo-mechanical treatment (training). The effect stabilizes during the training cycles in the preferred orientations of the martensite. This results in residual stress levels in the SMA element, which apply the required minimum stress to reset the SMA element by itself. During the training process, the intrinsic two-way SMA element is deformed to its real yield strength in both the austenitic and the martensitic state. Due to the plastic deformation of the material, dislocations and existing grid defects in the material are formed and stabilized. This, in turn, ensures residual stresses, which do not get degraded, not even in later phase transitions. After about 20–30 training cycles, a stable intrinsic two-way effect can be observed [2]. Due to lack of stability and a much smaller general effect compared to the extrinsic two-way effect, the intrinsic two-way effect is not used in actuator applications.

3.2.4 The Pseudo-Elastic Effect

Pseudo-elastic (or super-elastic) shape memory alloys make up the largest part of the shape memory technology market. In contrast to the one-way and two-way effect, the pseudo-elastic effect can be achieved mechanically at a constant temperature. This effect occurs at operating temperatures above the martensitic start temperature (M_S). Currently, obtainable pseudo-elastic shape memory alloys are usually completely austenitic at room temperature. Due to external mechanical stresses, the austenitic atomic grid gets deformed and a directly stress-induced detwinned martensite grid arises. Since this detwinned martensite is not thermodynamically stable, the grid retransforms into an austenite state upon mechanical relief. Figure 3.6 shows a

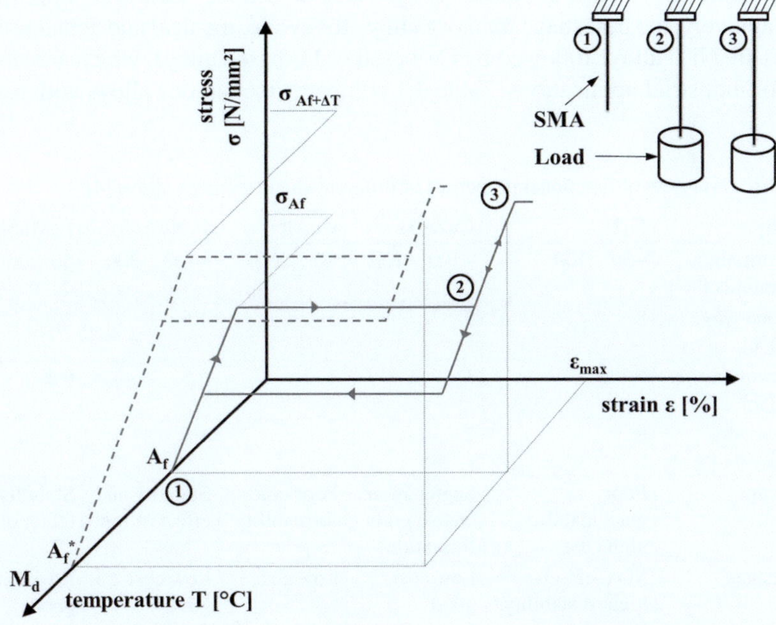

Fig. 3.6 Outline of the SMA pseudo-elastic effect

pseudo-elastic shape memory wire as main figure as well as a functional diagram. If loads are applied to pseudo-elastic elements, the material becomes elongated. Upon alloy relief, the pseudo-elastic wire reassumes its original shape. Figure 3.6 also shows a stress–strain diagram of a pseudo-elastic alloy with reference points to the main diagram.

If the material is mechanically loaded, it reaches a state in which the material deformation increases at nearly constant mechanical stress, namely after an approximately linear stress–strain deformation has occurred. This so-called stress plateau region of the curve allows for highly flexible deformations. When the stress-induced transformation is complete, the stress–strain curve rises again in a linear fashion. The exact amount of plateau stress depends on material temperature. This can be altered by thermo-mechanical behavior. With rising material temperatures, more mechanical energy must be introduced in order to initiate a phase transformation. Overall, the elongation of pseudo-elastic shape memory alloys can reach up to 8 % of their initial length. In comparison, the reversible spring-steel elastic yield strength lies approximately at 0.5 %.

3.3 Shape Memory Alloy Types

All alloy systems with a thermo-elastic martensitic transformation are conceivable, in principle, for the use as a shape memory alloys. Due to adverse mechanical properties, however, many of these alloys are of minor technical importance. Comparatively, the best shape memory alloys for cyclic, medical and actuator applications are NiTi alloys followed by Cu-based or Fe-based alloys, which are suitable only for industrial applications. Table 3.1 compares the various alloys with regards

Table 3.1 Overview of functional properties of different shape memory alloys [4]

Properties	NiTi	CuZnAl	CuAlNi	FeNiCoTi	FeMnSi
Transformation temperatures [°C]	−50…100	−100…100	80…200	−150…300	50…250
Max. one-way effect [%]	8	5	5	1,5	2,0
Max. two-way effect [%]	6	1	1	0,5	0,3
Max. pseudo-elasticity [%]	8	2	2	1,5	1,5
Handicaps	Poor machinability, high costs	Segregation, coarse grain formation	Poor cold formability	Stability and effect of low	Stability and effect of low
Advantages	Max. effects, highest stability, corrosion-resistant	Low cost, good formability	Low cost	Low cost, good formability	Low cost, good formability

to size of their utilizable shape memory effects, which are, in descending order, NiTi-based, Cu-based, and Fe-based alloys.

The material demands for industrial applications of SMA (such as valves) are as follows:

- High pseudo-plastic strain.
- Low deformation resistance of the martensite.
- High strength of the austenite.
- Good machinability of the material.

Additionally, the requirements for controlled elements in valve applications can be satisfied by the following demands:

- Good long-term stability of the shape memory effect.
- High conversion rate of electrical energy into mechanical work.
- High transition temperatures.
- Low thermal hysteresis for thermal-proportional control valves.
- Wide thermal hysteresis for thermal safety valves.
- Significant change of electrical resistance for electrical-proportional control valves.

To meet these requirements, and to use an actuator as efficiently as possible, only the use of NiTi alloys is recommended. This is nowadays encouraged by the general availability of NiTi materials. Besides the greatest shape memory effects, they also offer long-term stability and a long lifetime. With NiTi alloys, work cycles up to a million can be realized, however, to achieve such a high lifetime performance, a specific and detailed design of the shape memory components is essential [5].

3.3.1 Binary Nickel-Titanium Alloys

The most commonly used shape memory alloys are binary nickel-titanium alloys. The alloy system exhibits a high dependence on phase transition temperatures of its chemical composition. Thus, Fig. 3.7 represents the change in the martensite start temperature (M_S) when nickel content is added to the function (listed in atomic percent). When the chemical composition gets modified, for example, by adding 0.1 at.% of nickel content, the M_S temperature shifts by about 10 Kelvin. What is striking is the characteristic inflection point at about 50.5 at.% nickel. A reduction of the nickel content, and thus a substitution of the more expensive titanium, below this value affects only small increases in the phase transformation temperatures. This leads to high demands for the equipment used in the melting process. Sometimes the material needs to be re-melted several times to ensure a homogeneous composition of the ingot.

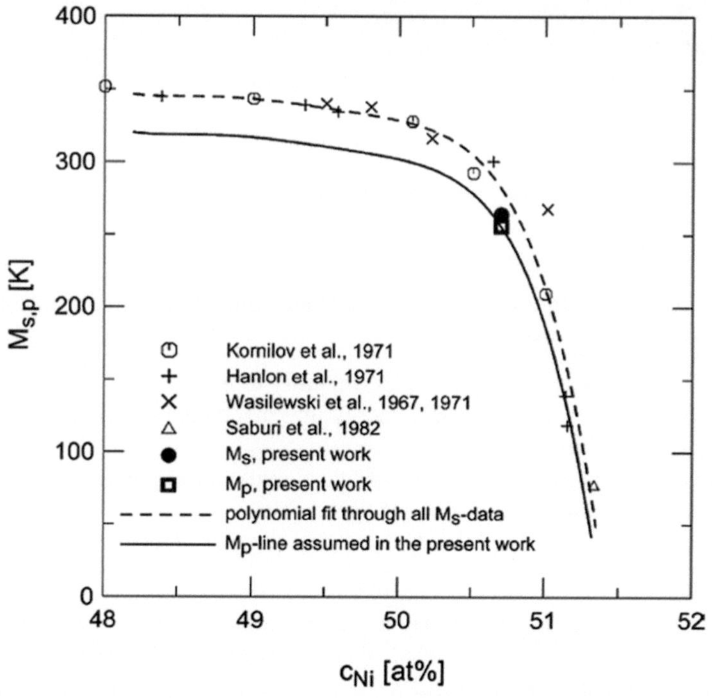

Fig. 3.7 Martensite start temperature as a function of the nickel component [6]

Table 3.2 The effect of ternary and quaternary elements in NiTi-based SMA [4]

Properties	C	O	N	H	Cu	Cr	Co	Fe	V	Nb
Transformation temperatures	↓	↓	↓	↓	→	↓	↓	↓	↓	→
Max. applied stress	↑	↑	↑	↑	↓	↑	↑	↑	↑	↑
Brittleness	↓	↓	↓	↓	→	↓	↓	↑	↓	→

3.3.2 *Ternary and Quaternary Nickel-Titanium Alloys*

With the addition of other elements to nickel-titanium alloys, the shape memory effect properties—such as phase transition temperature (PTT), hysteresis, and mechanical properties like strength and impact brittleness—can be modified. Table 3.2 depicts the qualitative influence of alloying elements on NiTi-alloys.

In addition, there are different ternary alloys, which have been researched and market-tested, containing elements such as hafnium, copper, niobium, palladium, platinum, or zirconium. For actuators, the NiTiCu alloy is most relevant. Available NiTiCu alloys include about 10 at.% of copper [2] and are used as substitution for a correspondingly high atomic percent share of nickel. NiTiCu alloys, however, exhibit considerable differences compared to binary NiTi alloys. On the one hand,

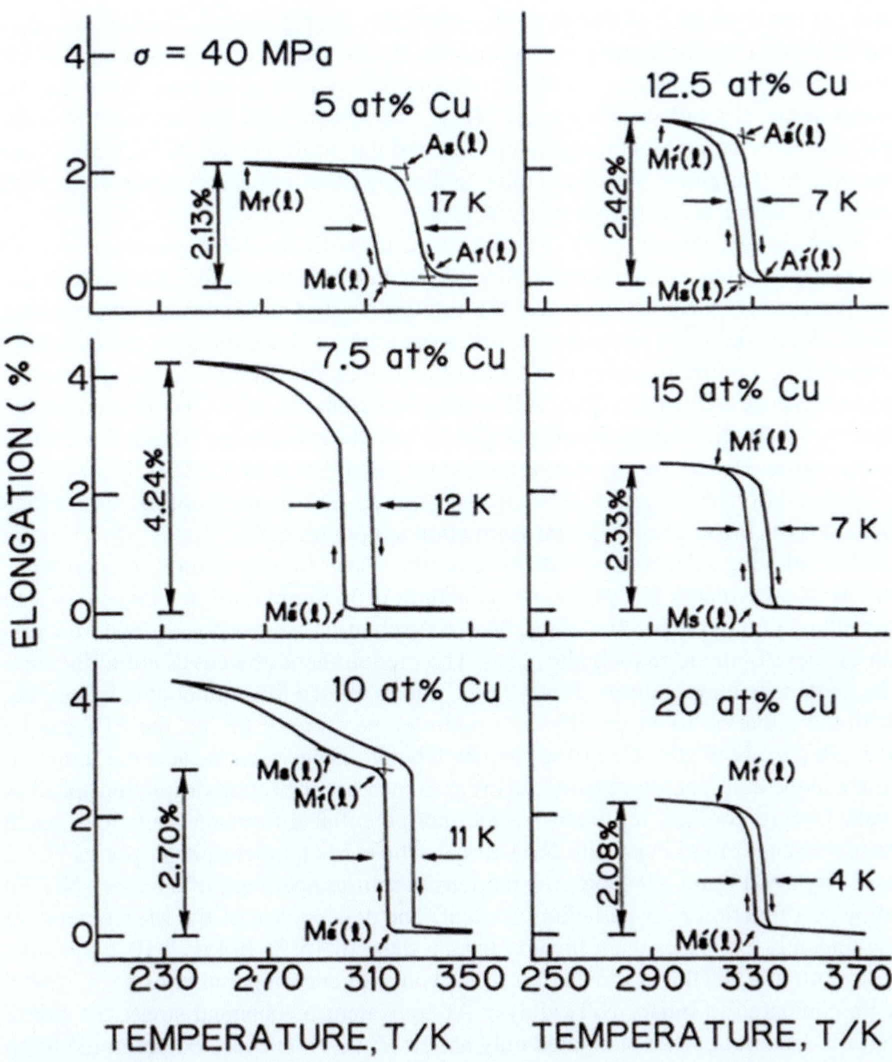

Fig. 3.8 Transformation behavior of NiTiCu depending on the amount of copper [7]

the transformation hysteresis is considerably narrower than with binary alloys, but only at the cost of reducing the mechanical output potential. Thus, NiTiCu alloys can only permanently generate about half as high mechanical tensile stresses in comparison to NiTi alloys. Figure 3.8 shows the variation of material hysteresis NiTiCu of alloys as a function over copper content. In this figure, the elongation is reset by thermal heating. The gradient of the hysteresis can be significantly higher, and can consist of several gradient sections as displayed in Fig. 3.8 (for example at 10 % for Cu). A striking advantage of this alloy type is better fatigue performance and the linear change of electrical characteristics during electrical actuation, compared to NiTi. Unfortunately, the application field of NiTiCu alloys is limited to the lower ambient temperature field considering electrical applications. Since the M_F temperature can only reach around 50 °C, the system will not retransform completely when the ambient temperatures exceed this limit. Hence, NiTiCu alloys are not suitable for automotives, aircrafts, and many other industrial applications with ambient thermal fields higher than 50 °C.

Interestingly, ternary NiTi (X) alloys, and in particular NiTi quaternary (X, Y) alloys, are known as high temperature alloys. In these alloys, the martensite finish temperature (correspondingly also the martensite start temperature) reaches over 80 °C. Industrial users and potential industrial users of shape memory alloys require these types. Common application fields of actuators, for example in the automotive and aerospace industry, require full system functionality at 85 °C. This equals an electric system that retransforms completely into the martensitic state at this critical temperature. Hence, the M_F temperature must be higher than 85 °C.

A similarly high system dynamic is required due to the cooling rate, which demands the highest possible transformation temperature.

The alloying additives, which increase the phase transformation temperatures, are usually elements like platinum, palladium or hafnium, and in some cases also tantalum–aluminum combinations. Material-scientific research worldwide focuses on the development of such alloys [8]. The predominant objectives are to increase the martensite temperatures above 100 °C and to reduce the temperature hysteresis, with the intention to ensure that at a room temperature of 85 °C, the PTT can be increased to about 200 °C, will say to a high potential between the actuator temperature and the ambient temperature, in order to increase the retransformation speed as well. During cooling, an extensive thermal-potential difference leads to a much faster retransformation as with NiTi alloys, whose Ms temperature surpasses 80 °C only slightly. Figure 3.9 shows the martensite start temperature of a ternary NiTiPd alloy as a function over palladium content. The dependence of the M_S temperature significantly expresses itself linearly from a share of 15 %. In Fig. 3.10, a dynamic test result of a NiTiPd is shown over time at an ambient temperature of 85 °C (SMA wire compared to binary NiTi alloys). At equivalent mechanical stress, the brittle NiTiPd material can be elongated only about 1.5 %. NiTi could be deformed to up to 4.4 % in this experiment.

Both samples (diameter: 0.2 mm, length: 120 mm) are activated by a heating current of 1 A for a heating time of 2 s. Comparing the two experiments shows that the NiTiPd alloy transforms more slowly. This is due to the longer necessary heating time to reach the A_F temperature, which is higher compared to the conventional NiTi at about 100 °C. Taking cooling into consideration, the figure also shows that a NiTiPd alloy retransforms significantly faster than a binary NiTi sample.

However, the mechanical properties of this material allow only for small distortions, and therefore further technological breakthroughs are needed to achieve proper usability. Ultimately, the use of this material in the automotive and automation industry is questionable from an economic perspective. Since the price of palladium equals 1,500 times the price of nickel and titanium, cost-per-kg for high temperature

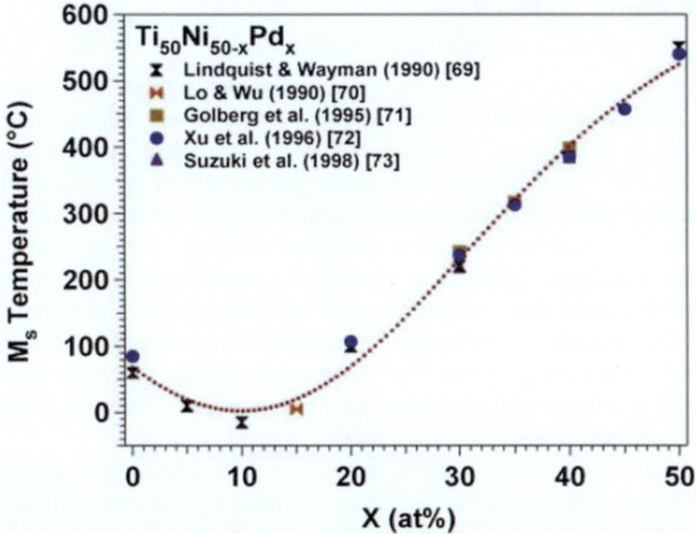

Fig. 3.9 PTT dependence of palladium content in NiTiPd alloy [8]

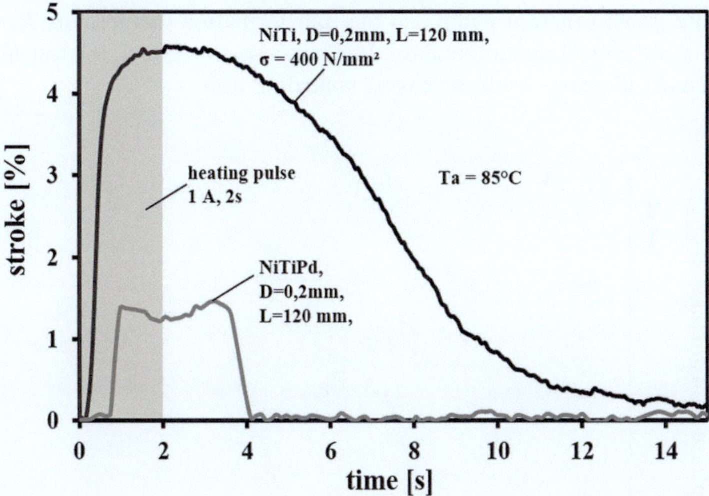

Fig. 3.10 Comparison of dynamic properties of NiTi and NiTiPd alloys in comparable conditions

shape memory alloys with a 30 % share of palladium would economically not be feasible for many industrial applications. Hope is placed on NiTaAl alloys [9], because this alloy combination is more economical, and thus production has signifi-cant advantages over NiTiPd, NiTiHf and NiTiPt systems at high-phase transition temperatures (approx. 130 ° C).

3.3.3 R-Phase NiTi Alloys

R-phase transformation (R) precedes the transformation from austenite (B2 resp. A-phase) to martensite (B19 resp. M-phase) in thermo-mechanically treated near-equiatomic NiTi alloys, in Ni-rich NiTi alloys aged at an appropriate temperature between 300 °C and 550 °C, and also in ternary NiTiFe or NiTiAl alloys [10]. The appearance of R-phase transformation can be explained through lattice defects in microstructures such as dislocations or precipitations [11]. In contrast to transformation from martensite to austenite (up to 8 %), transformation from austenite to R-phase yields a small shape change (up to 1 %) and generates less power. The typical temperature hysteresis during transformation from austenite to the R-phase is very small (2–5 K) compared to the transformation from austenite to martensite (20–70 K) [12]. Figure 3.11 compares the presented transformations using a wire actuator as example.

The development of shape memory materials and the handling of actuators based on the R-phase transformation are very complicated. Factors influencing the R-phase transformation negatively are: false material development (e.g. thermo-mechanical pre-treatment), incorrect actuator design (e.g. false shear strain) or misuse during the production (e.g. overstress). Figure 3.11 shows important influences on selected properties of R-phase actuators for Ni-rich NiTi alloys. Rising values of some influences cause different and sometimes opposite effects. For example, the temperature transformation width and the transformation temperature R_S decrease with increasing annealing temperature. In contrast to this effect, the transformation temperature R_S increases with increasing annealing time.

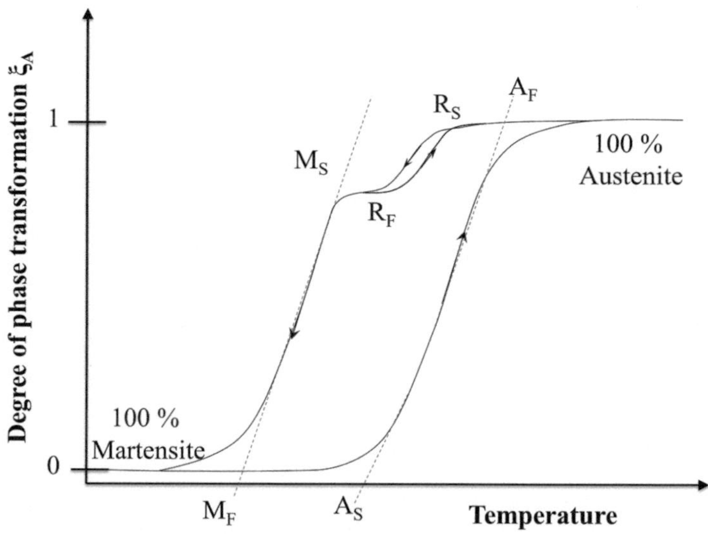

Fig. 3.11 The characteristics of a pre-stressed SMA with R-phase transformation

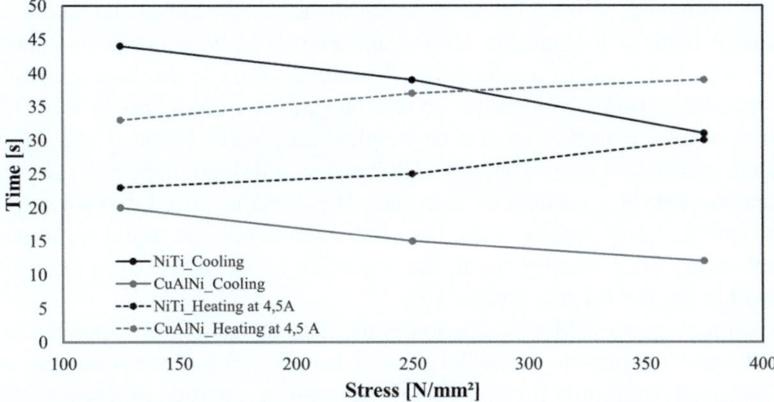

Fig. 3.12 Comparison of dynamic properties of CuAlNi alloy to binary NiTi alloy in comparable testing conditions

3.3.4   Copper-Based Alloys

The best-known copper-based alloys are CuZnAl and CuAlNi. They cannot be used as implant materials due to their chemical composition and are therefore suitable only for industrial purposes. While CuZnAl enables only low-phase transformation temperatures (for industrial applications), CuAlNi reaches in certain chemical compositions even the high temperature needs of the automotive industry. Alloys containing a weight percentage of about 12 % aluminum and 5 % nickel have an M_F temperature of about 120 °C. Mechanical characterizations further show reversible and high mechanical stresses [13]. The maximal deformation (also contraction) can reach 8 % as well as high mechanical stresses of about 350 N/mm². A major disadvantage is the processing ability and the brittle behavior of these materials. The wire drawing process is sometimes not suitable for these materials, which is why alternative design methods are used. Regarding actuator activation, more problems are noticeable. Due to the low electrical resistance of copper, high temperature copper-based SMA need much higher electrical activation currents than comparable NiTi elements (Fig. 3.12).

3.4   Manufacturing of Shape Memory Alloys

In order to understand the influences on SMA material parameters, it is sometimes necessary for product developers to understand the production processes of the material itself. In this chapter, a brief summary of the production process from melt to an SMA round wire is given. For detailed data on production processes of SMA, [3] offers detailed explanations.

At the beginning, an ingot has to be produced out of pure nickel and pure titanium. A common method for melting NiTi shape memory alloys is vacuum induction melting, which is used to produce the SMA ingot. Due to the high sensibility of phase transformation temperatures to the Ni-ratio, as discussed in Fig. 3.6, the quantities of used materials have to be weighed accurately. Figure 3.13a shows primarily the melting process. Through high frequency induction, heated raw material components melt in a vacuum or inert gas. The mold is stirred and solved, which leads to perfect homogeneity of the material. Afterwards, the liquid is molded into an ingot form. After cooling down, the ingot can be extracted from the form and processed in the hot forging process (b).

The critical step in SMA production is the thermal treatment which has significant influence on plateau stress levels and transformation temperatures. Alloys which are used commonly for SMA electric actuation are often programmed by the SMA material suppliers for optimal behavior.

However, a cold-worked SMA material has to be programmed for its optimal behavior. Figure 3.14 shows in the left part the major influence factors on the heat treatment. All heat treatments have a connection to the SMA element's dimensions and its volume/mass which has to be heated. The duration of the heat treatment has also influences on the SMA characteristics as well as the treatment temperature. Also the atmosphere has significant influences, mainly on the surface of the SMA element (relevant for medical applications).

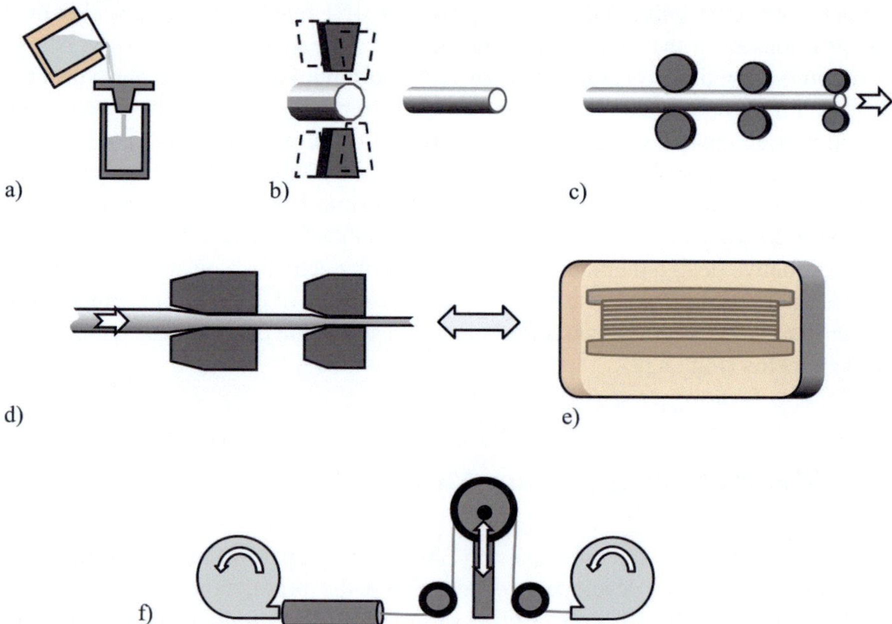

Fig. 3.13 The production process of SMA wires in various steps: a) vacuum-induced melting of NiTi, b) round forging in hot state, c) hot rolling, d) wire drawing, e) thermal treatment, f) thermo-mechanical training

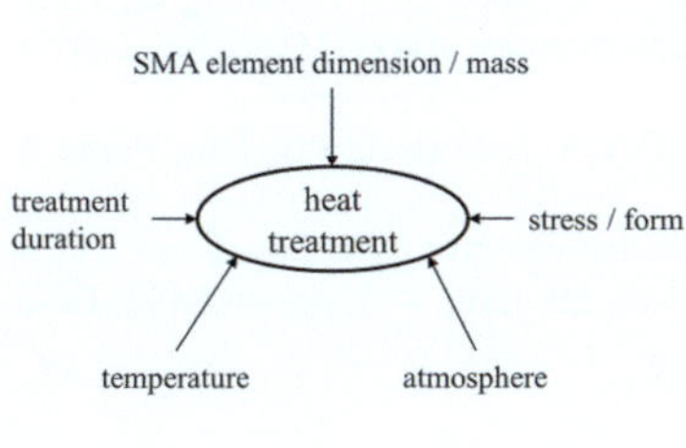

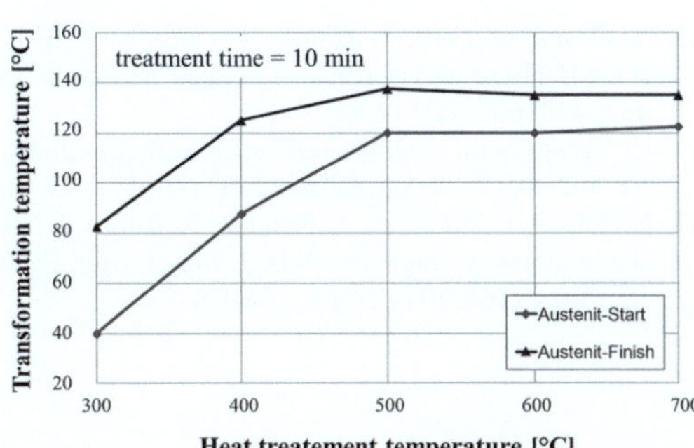

Fig. 3.14 Major influences on SMA heat treatment (left), experimental result of thermal heat treatment (right)

Normally, SMA straight wire elements are treated without stress, but also the imprint of the shape can be used by the utilization of a tool. In such case, the SMA element is heat-treated at a relative small mechanical stress level.

In the right part of Fig. 3.14, a result of heat treatment of a NiTi (54,8 wt. % Ni) SMA wire is presented. The austenitic temperatures increase with heat treatment temperatures up to 138 °C. The experimental validation has been made at stress levels higher than 350 N/mm^2.

References

1. H. Stork, *Aufbau, Modellbildung und Regelung von Formgedächtnisaktorsystemen.* VDI-Fortschrittsberichte, Reihe 8, Nr. 657. (VDI-Verlag, Düsseldorf, 1997)
2. M. Mertmann: *NiTi-Formgedächtnislegierungen für Aktoren der Greifertechnik.* VDI-Fortschrittsberichte, Reihe 5, Nr. 469 (VDI-Verlag, Düsseldorf, 1997)
3. K. Otsuka, C.M. Wayman, *Shape Memory Materials* (Cambridge University Press, Cambridge, 1998)
4. S. Langbein, A. Czechowicz, *Konstruktionspraxis Formgedächtnistechnik* (Springer Vieweg, Mannheim, 2013). 3834819573
5. D. Stöckel, *Engineering Aspects of Shape Memory Alloys*, 1st edn. (Butterworth-Heinemann Ltd., London, 1990)
6. J. Khalil-Allafi, A. Dlouhy, G. Eggeler, Ni_4Ti_3-precipitation during aging of NiTi shape memory alloys and its influence on martensitic phase transformations. Acta Mater. **50**, 4255–4274 (2002)
7. K. Otsuka, X. Ren, *Martensitic transformations in nonferrous shape memory alloys.* Materials Science and Engineering, A 273–275, Elsevier, 1999, S.89–105
8. J. Ma, I. Karaman, R.D. Noebe, *High temperature shape memory alloys.* Int. Mater. Rev. **55**(5), 257–315 (2010). Maney, ASM International
9. S. Miyazaki et al., Novel beta-TiTaAl alloys with excellent cold workability and a stable high-temperature shape memory effect. Scr. Mater. **64**(12), 1114–1117 (2011). Elsevier Science, Oxford
10. M.-S. Choi, J. Ogawa, T. Fukuda, T. Kakeshita, *Stability of the B2-type structure and R-phase transformation behavior of Fe or Co doped Ti–Ni alloys.* Mater. Sci. Eng. A **438–440**, 527–530 (2006). Elsevier

11. Y. Zhoua, G. Fana, J. Zhang, X. Ding, X. Ren, J. Suna, K. Otsuka, Understanding of multi-stage R-phase transformation in aged Ni-rich Ti–Ni shape memory alloys. Mater. Sci. Eng. A **438–440**, 602–607 (2006)
12. D. Treppmann, *Thermomechanische Behandlung von NiTi.* VDI-Fortschrittbericht, Reihe 5, Nr. 462, (VDI-Verlag, Düsseldorf, 1997)
13. S. Pulnev, V. Nikolaev, A. Priadko, A. Rogov, I. Vahhi, Actuators and drivers based on CuAlNi shape memory single crystals. J. Mater. Eng. Perform. Vol. 20, Issue 4–5, pp. 497-499, DOI: 10.1007/s11665-011-9915-2 (2010)

Chapter 4
Introduction to Shape Memory Alloy Actuators

Sven Langbein and Alexander Czechowicz

4.1 General Overview of SMA Actuators

For the use of SMA in valve systems, commonly the extrinsic two-way effect is used due to its high mechanic output. Generally, the effect can be further distinguished into two different kinds of applications: electrical and thermal. In Chap. 6, this distinction will be discussed in more detail and for different application types. To help understanding SMA, a simple thermal valve model is shown in Fig. 4.1. The SMA element is pre-stressed by the main valve piston, which is linked to a conventional spring element. The fluid enters the chamber that contains the SMA element and the thermal field of the fluid is transported to the SMA material itself, causing the phase transformation from martensite to austenite. This is linked to a macroscopic change of the shape of the spring form. While the imprinted shape is a long spring, the material changes into the deformed (compressed) shape using the heating energy of the fluid. On the right-hand side, the displacement of an SMA spring is shown schematically over temperature. There are different types of thermal SMA characteristics. The hysteresis can be wider, narrower or displaced to lower or higher temperatures. The gradients of the transformation curves as well as the general transformation behavior are a function of the material parameters: material production, heat treatments, and thermo-mechanical training. As presented before, the R-phase NiTi has a much lower hysteresis, but also a lower mechanical potential in comparison to conventional NiTi alloys.

S. Langbein (✉)
FG-INNOVATION GmbH, Universitätsstr. 142, 44799 Bochum, Germany
e-mail: langbein@gmx.com

A. Czechowicz
Zentrum für Angewandte Formgedächtnistechnik, Forschungsgemeinschaft Werkzeuge und Werkstoffe e.V., Papenberger Straße 49, 42859 Remscheid, Germany
e-mail: czechowicz@fgw.de

© Springer International Publishing Switzerland 2015
A. Czechowicz, S. Langbein (eds.), *Shape Memory Alloy Valves*,
DOI 10.1007/978-3-319-19081-5_4

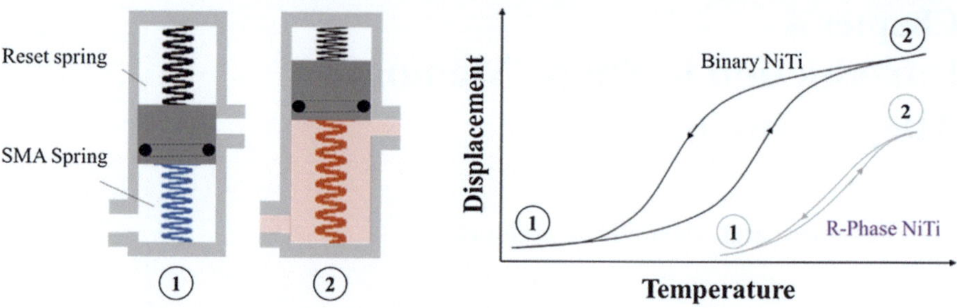

Fig. 4.1 Functional characteristics of thermal SMA actuators in valve applications

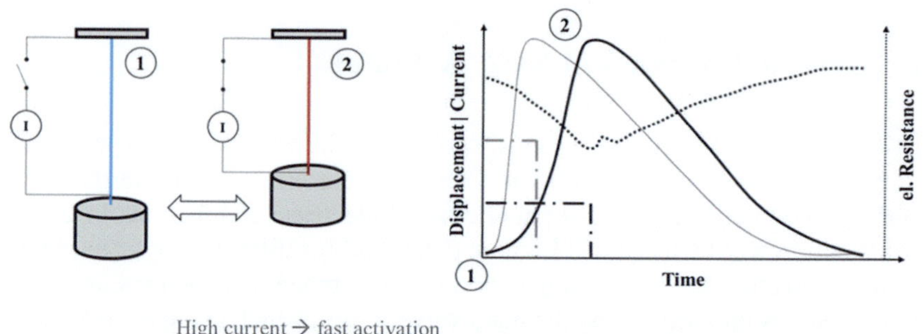

Fig. 4.2 Functional characteristics of an electrified SMA actuator

The dynamic behavior of an electric SMA actuator resembles, at first sight, a simple linear actuator. If the SMA wire, as shown in Fig. 4.2, is loaded with a mechanical pre-stress (in this example with a constant mass), it is elongated by up to 8 % of its original length. In usable cases, this pre-strain is one of the dominant factors for lifetime performance. Heating, especially electrical activation is a dynamic process. During this action, the properties of the SMA significantly change due to the described phase transformation. This process is influenced by a large number of functional factors, which have to be considered during the engineering process. On the right-hand side of Fig. 4.2, a dynamic test is shown schematically.

On the left Y-axis, displacement and electrical current are noted, while the right Y-axis normally indicates electrical resistance. The X-axis shows time. If an SMA wire is triggered with a high electric current, the transformation from martensite to austenite is faster than with a lower electric pulse. Therefore, the lower current level has to be upheld for a longer period of time. It should be noted that after electrical deactivation, the retransformation speed from austenite to martensite has the same gradient. Hence, the electrical activation has in this case hardly any influence on the deactivation speed of the SMA actuator. Unfortunately, this is not always the case. If the pulse time is unsuitable, the system will be overheated. This results in an earlier fatigue and longer deactivation times. A very important characteristic of

SMA actuators is the change in electrical resistance shown in Fig. 4.2. The change from martensite to austenite triggers a decrease in electrical resistance. Austenite has a much lower specific resistance level, which can be seen during the phase transformation of mixed austenite–martensite phases. Details on resistance behavior, especially on potential for valves as controlled devices, will be given in Chap. 5. The behavior of shape memory actuators is determined by numerous actuator properties that are already defined in the development process. Thus, the diameter of an SMA actuator affects not only the maximum pulling force, but also the dynamic heat-up and cool-down properties. The applied mechanical bias (often the same as the tensile force) changes the phase transition temperatures and the maximal stroke. Figure 4.3 shows the functional relationships of an SMA actuator exemplified by a shape memory wire [1]. All variables must be set in the design of an SMA-driven valve. These variables do not only change with the material-intrinsic function of the wire temperature, but also its cyclic fatigue.

The main parameters can be defined as follows:

- Heating speed: The time period in which a shape-memory actuator experiences a change in temperature from the ambient temperature to the A_F temperature.
- Cooling speed (also retransformation time): The time in which an actuator is cooled down to M_F temperature. This time can be divided into sections of inactive cooling time (t_{ia}) and active retransformation time (t_a). The inactive cooling time describes the cooling time from the end of energy input into the system, which may be delayed due to overheating of the material, until the active retransforma-

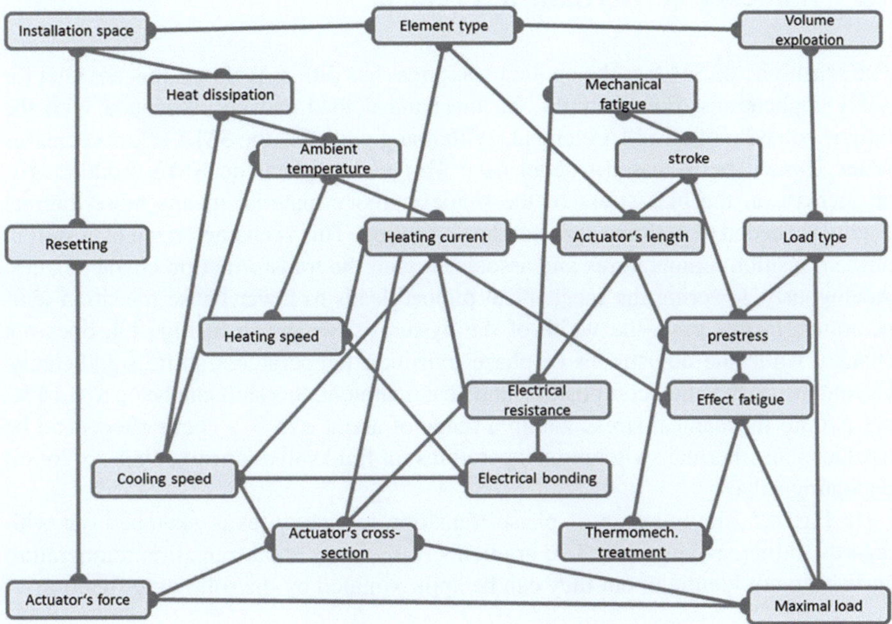

Fig. 4.3 Network of influences on SMA wires under preload [1]

tion can start at A_F temperature. During the active cooling time, the visible decrease of stroke can be observed in shape memory actuators.

- Ambient temperature: Defined as the temperature of the medium (solid, fluid or gas) that directly surrounds the shape memory actuator.
- Resetting: Needed for the use of the extrinsic two-way effect. While in use, the resetting can be a constant mass, a linear return spring, or a nonlinear pseudo-elastic member element.
- Stroke (also contraction): The resetting of the elongation, which is caused by stress induced by the resetting element.
- Actuating force: Mechanical stress that is generated by the phase transformation due to the material contraction, calculated over the square area of a shape memory element.
- Variance of the phase transition temperatures (PTT): Phase transition temperatures depend on mechanical load. With increasing mechanical stress, the transformation temperatures increase almost linearly. Figure 4.5 shows an experimental result in a climatic chamber in which a shape memory wire was measured under different loads. The mechanical, conversion-characteristic dependence of the phase transition temperature is listed as function of the applied mechanical stress. This diagram helps reading out the use of present shape memory actuators in a thermal field. The austenite temperatures range between 80 °C and 130 °C, while the martensite temperatures are lowered by the hysteresis width.

4.2 Influence of Mechanical Preload

The sensitivity of SMA to the applied bias stress has different effects that are vital for SMA applications. For example, the mechanical load can be associated with the fatigue behavior of an SMA element. With rising preloads, the SMA effect decreases faster. This will be discussed in detail in the "Fatigue" chapter under SMA. Additionally, an increase in the bias stress of the shape memory material means more thermal energy is needed to perform a phase transformation. This is characterized by a shift in phase transition temperatures and associated with the transformation characteristics. Analogously, lowering the mechanical preload leads to lower phase transition temperatures. In this case, the width of the hysteresis, as shown in Fig. 4.4, does not change, while the adjustment of phase transition temperatures shifts significantly. Depending on the alloy composition and heat treatment, this shift can be up to 0.14 °C per N/mm² mechanical stress within a range of about ±25 °C. These effects can be used to adjust thermal switching temperatures for fluid valves as mixer taps, or for oil regulating valves.

In Fig. 4.5, the increase of phase transition temperatures is sketched out with regards to increasing stress. The gradients of the individual transition temperature curves are not identical, but they can be approximated by the following function:

$$dT/ds\sigma = 0.11 \text{ to } 0.14 \left[(°C)/\left(N/\text{mm}^2 \right) \right]$$

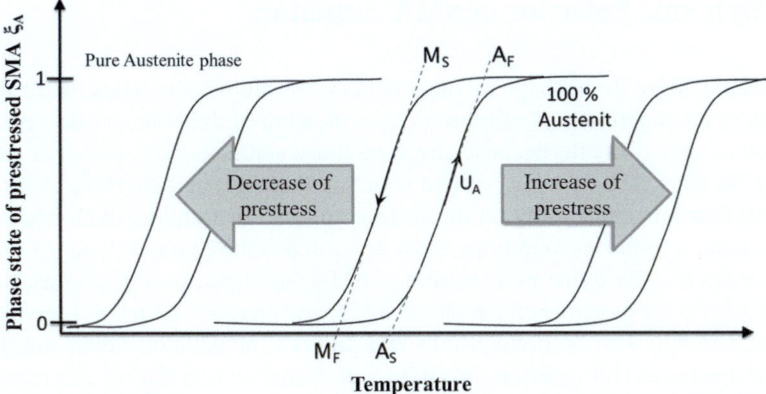

Fig. 4.4 Variations in pre-stress of an SMA actuator resulting in changes to the transformation hysteresis

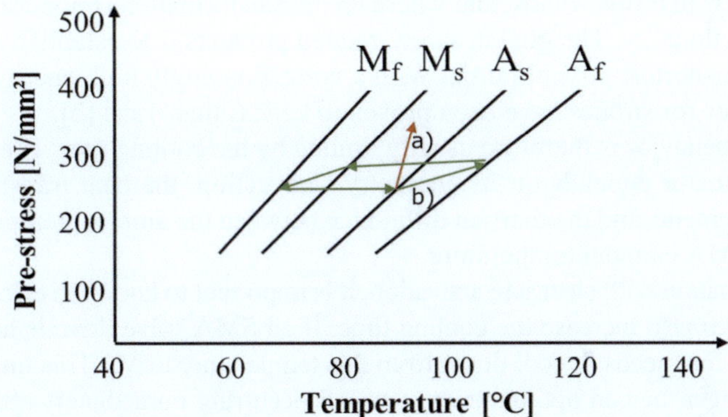

Fig. 4.5 Variance of phase transition temperatures with increasing mechanical stress

Consequently, the response time of an SMA element decreases with increasing load within constant activation parameters. Thus, the same shape memory element retransforms faster when the mechanical preload is increased. The significance of this change depends on the characteristics of the linear rising load. This should be considered especially during the engineering process. If a shape memory valve works against a linear, increasing load, the transformation temperatures will increase too. If a system starts opening at 250 N/mm^2, pre-stress at 87 °C, with a rising bias characteristic, the transformation could still be stopped by stable thermal fields and fast rising pre-stress up to 350 N/mm^2 at only 95 °C (see path a) in Fig. 4.5). In this case, the system would not activate properly. Hence, with thermal SMA valve drive design, the proper characteristic of the resetting element is crucial. Following path b) in Fig. 4.5, an SMA valve opens at 87 °C. The thermal shift caused by lesser rise of mechanical preload to 280 N/mm^2 enables a full valve opening at 105 °C. The retransformation starts at this mechanical stress level at 79 °C and ends at 65 °C, with a stress level of 250 N/mm^2 once again.

4.3 Dynamic Behavior of SMA Actuators

The dynamic behavior of shape memory actuators is highly dependent on the design parameters and operating conditions. A classification of dynamic application areas is therefore very difficult, because shape memory actuators with different geometries can be used. The dynamic system response (heating) and the retransformation (cooling) depend significantly on the surface volume. In addition, exploited effects of heat transfer to the surroundings, such as a forced convection or heat-dissipating media, influence the dynamic behavior of SMA actuators. In this chapter, a brief introduction to the dynamic characteristics of shape memory actuators is given. The dynamic behavior can be divided into two different application fields: single use reaction dynamics (for example in safety valves that switch during an emergency incident) and cyclical dynamics (for example for valves in dosing applications).

While the single use reaction behavior is only characterized by the heating process, the cyclical dynamics depend mainly on the cooling behavior. An analysis of the reaction behavior is vital for safety applications where actuators have to be triggered quickly, in milliseconds, and where the retransformation is noncritical for the product functionality. The goal in safety-related products is to establish a force or a stroke in the shortest possible time. With a correspondingly high energy input, the reaction times for strokes have been proven to be less than 1 ms [2].

Cyclical behavior is therefore usually limited by the cooling rate. The cooling of an SMA actuator depends on its geometry, convection, the heat transfer into the bounding elements and the thermal difference between the ambient temperature and the actual SMA element temperature.

In combination with electrical activation, it is important to know the effects of overheating, which also increase the cooling time. If an SMA valve drive is heated up to $A_F + 50$ K, it first needs to cool down from this temperature to M_s. This time period is of course longer than an optimal energy cut-off occurring immediately after reaching austenite finish temperature. The rate of heating and cooling depends on various factors. The cooling processes are often slower than the heating processes in macroscopic electric applications because the heating speed can be controlled by electrical current. However, by downscaling the elements, the material volume (a cubic function) reduces greatly. Because the downscaling of the surface is based only on a quadratic function, the ration of surface area to material volume changes massively.

As a result, the gradient of heat transfer and convection increase results in faster dynamics. With the use of an SMA thin wire, valves and camera systems can reach working frequencies at more than 10 Hz Actuator 2012, Cambridge Mech, limiting the application mainly to force output. Still, this faster dynamic is influenced by shifting ambient conditions.

The ambient conditions cannot be ignored when considering the dynamic behavior. Ambient conditions are on the one hand thermal fields, but on the other hand constructional and environmental parameters such as working medium or a forced convection (by wind for example). An SMA element in an air-permeable casing cools faster through air-flow than an element in a capsule casing.

Influences on cooling behavior are shown in Fig. 4.6. Cooling behavior can be influenced by five main factors which are the transformation temperatures, the hysteresis width, the ambient conditions, the SMA element's geometry and the control strategy.

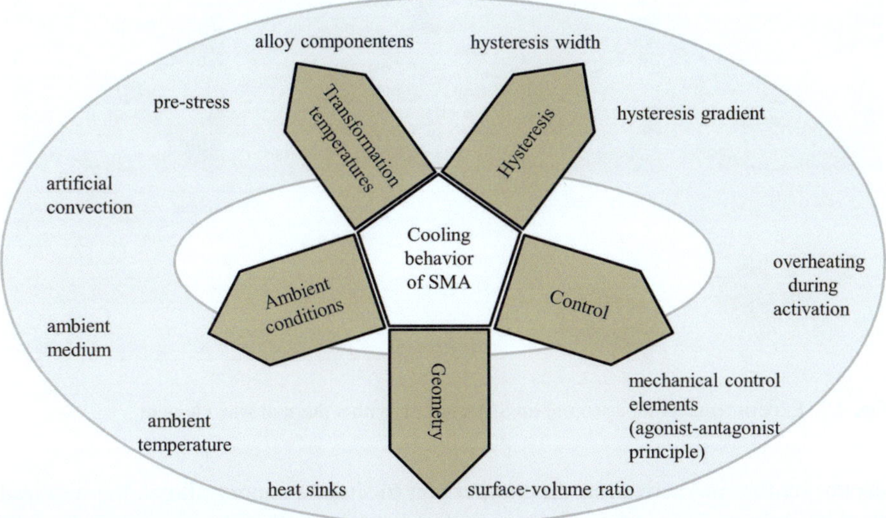

Fig. 4.6 Factors influencing cooling behavior

Another major effect on dynamic behavior is the SMA element's shape. Forms such as coil springs, for example, need more heat for activation due to the larger cross sections needed compared to wires with the same force output, but at greater ratio of element dimension and stroke. Additionally, inhomogeneous material stress in spring design leads to varying transformation temperature levels along the spring geometry. Hence, mechanical design of spring elements has an important influence on the thermal (and with it the electrical) dynamic characteristic.

The hysteresis form has a big impact on the dynamics of SMA elements. The form can be generally described by the hysteresis width and the transformation gradients. An actuator with a large hysteresis has to be cooled to low temperatures in order to start retransformation. The time between the deactivation at A_F and reaching M_s, as discussed above is the reason for the so-called "dead-time" before the retransformation starts. There is also a dead-time at the end of the transformation. Hence, the heat conduction and inhomogeneity must be minimized in order to improve the dynamics. Small hysteresis widths are therefore advantageous for applications with high and precise dynamics, while large hysteresis SMA can be used for safety functions with a thermal delay.

4.3.1 Comparison of Cyclical Dynamics in Thermal Applications

Nowadays, many thermal switching and controlling elements are using wax expansion elements filled with thermo-sensitive liquids. These liquids are alcohols or thermo-sensitive waxes that expand their volume with rising temperatures. A pre-stressed device filled with such liquid is able to generate high forces and moderate

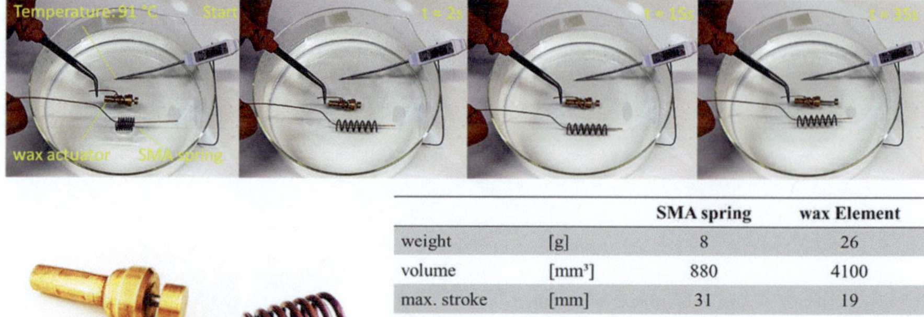

		SMA spring	wax Element
weight	[g]	8	26
volume	[mm³]	880	4100
max. stroke	[mm]	31	19
max. Force	[N]	37	40
act. temperature	[°C]	$A_s = 75 / A_f = 86$	start 78 / finish 85

Fig. 4.7 Experimental comparison of an SMA spring with a thermal wax element

strokes by thermal activation. In comparison to shape memory alloys, the required element volume is significantly higher, because the expansion element's liquid's state change has a much lower effect compared to SMA. In Fig. 4.7, a simple comparison has been drawn in nearly similar thermal conditions between an expansion element and an SMA spring actuator. The prepared water had an initial temperature of 91 °C. Simultaneously, both actuators have been inserted into the thermal bath without load. Because the SMA's volume is nearly 21 % of the expansion element's volume, the thermal gradient within the SMA spring rises immediately to the austenite start temperature. The austenite finish temperature is reached in less than 2 s. It should be noted that the phase transition temperatures are slightly lower than the fluid temperature. With rising difference between the fluid and the SMA's phase transition temperatures, the thermal gradient would rise and with it the transformation speed. Compared to this experimental result, the expansion element needs more than 30s to perform the same transformation. This is caused a) by the element volume and b) by the liquid, which has been used inside the expansion element.

4.3.2 Comparison of Cyclical Dynamics in Electric Applications

In order to assess the applicability of SMA in relation to electrical, cyclical dynamics, a simple comparison of SMA actuators (here just SMA wire forms), piezoelectric actuators, solenoids and hydraulic drives of comparable size for fine mechanical applications is shown in Fig. 4.8.

By limiting the maximum contraction of SMA, it is possible to achieve high numbers of cycles. Up to a certain total actuator length, the use of SMA wires no longer makes sense, because otherwise the component dimensions and complexity would exceed the benchmark of electro-dynamic actuator elements. Through various element designs (i.e. springs), the stroke can be increased so that the system

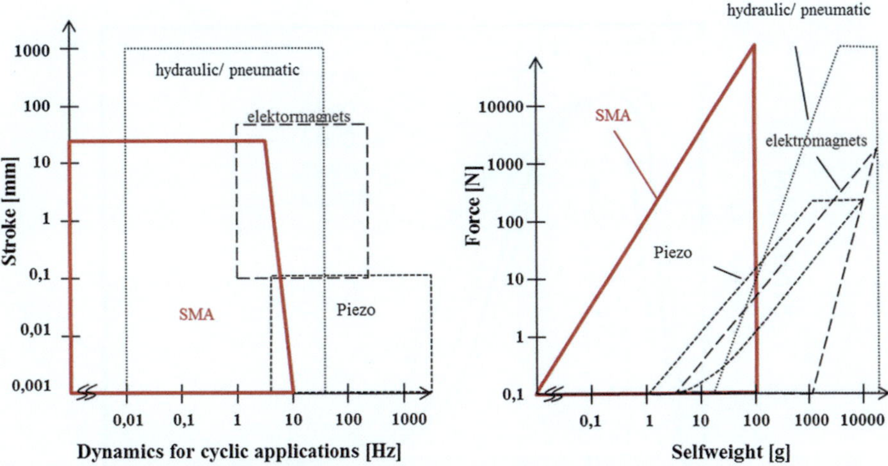

Fig. 4.8 Cyclic dynamics of shape memory wires (highlighted in *red*)

complexity, and thus the competitiveness with other actuation principles, will be reduced. Therefore, the adjustment path is shown as a horizontal line at about 20 mm travel range in the left part of Fig. 4.8. However, stroke has only a minor influence on the dynamics of SMA. Since phase transformation proceeds diffusion-less, it is not time-dependent. The amount of material to be transformed, and the travel, is thus not negatively affected by the dynamic balance. Due to the cooling processes, the cyclic operating frequency can reach up to 10 Hz; in some micro-system tests even up to 100 Hz. With smaller wire diameters, the cyclic frequencies for actuation rise. But smaller diameters equal smaller pulling forces (the pre-stress affects the SMA element diameter). Therefore, a decreasing actuator force is listed with increases in cyclic dynamics in Fig. 4.8 on the right. However, a slow and smooth transformation can be reached by using SMA and by having a closer look at the reaction dynamics. These can be adjusted much easier, unlike the cyclical dynamics. A variance of the current level, for example, can accelerate the activation of the shape memory element. However, it must be ensured that increased ambient temperatures do not lead to overheating of the element. Figure 4.9 shows the measurement result of a 0.2 mm thick shape memory wire (at a constant stress level of 200 N/mm^2), which was activated with different current levels. The control of the energization duration was realized with a resistance-based automatic shutdown system, which stopped the activation dependent on the gradient level of the stroke. If the stroke increase stopped, the current was switched off as well. In some of the curves, the input energy was not sufficient for a complete transformation up to 3 % stroke, which was adjusted by pre-strain before the experiment.

The activation of the actuator at a low current of 500 mA lasts almost four times as long as the fastest activation at 900 mA.

For clarity on how to apply the proper electric activation, it is advisable to introduce the electric current as current density J in amperes per square millimeter wire

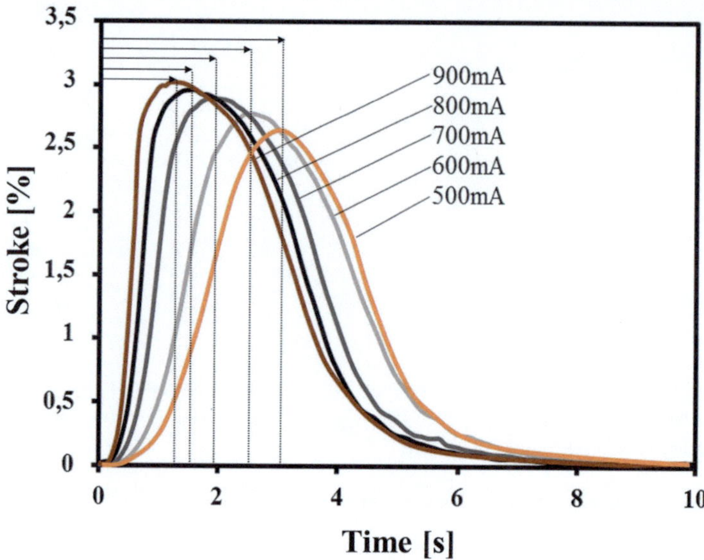

Fig. 4.9 Effects of variance in electric current on the reaction dynamics of a shape memory wire at 200 N/mm^2 load and with 0.2 mm diameter [3]

cross-section. This artificially generated variable is useful to estimate the reaction time as a function of the applied electric current for several wire sizes. Figure 4.10 shows a reaction time chart for shape memory wires with a cross-section between 0.15 and 0.25 mm. Various wire lengths between 100 and 300 mm were measured to possibly be able to formulate statements about variances that may occur due to length-dependent differences in resistance. Above the characteristic line is the selected current density with which the activating energy of the shape memory wire is too high. The risk of material damage caused by overheating is large in these parameters. Below the characteristic line, a damage is unlikely. However, a complete contraction of the SMA element, as presented in Fig. 4.9 at 500 mA, is not achieved. The time range for heating (<1 s) is particularly difficult to apply in this diagram. The required electrical energy rises to high voltage applications. Furthermore, it should be noted that Fig. 4.10 is a representation of specific functional parameters (load = 200 N/mm^2, ambient temperature = 20 °C). With any change to both parameters, the shown curve would move significantly. If the ambient temperature rises, a curve displacement to the left would be observed. If the mechanical bias stress is increased, the transformation temperatures of the shape memory element would also increase.

Thus, the characteristic shown in Fig. 4.10 would be shifted to the right. The same applies of course to the opposite changes in ambient temperature and bias stress. Lowering the ambient temperature would shift the curve to the right, whereas it would be displaced to the left by load reduction.

To make a more detailed indication of the dynamic activation of shape memory actuators, a dynamic overview diagram can be created. Such a diagram is shown in Fig. 4.11. Again, the diagram applies a constant stress of 200 N/mm^2. The

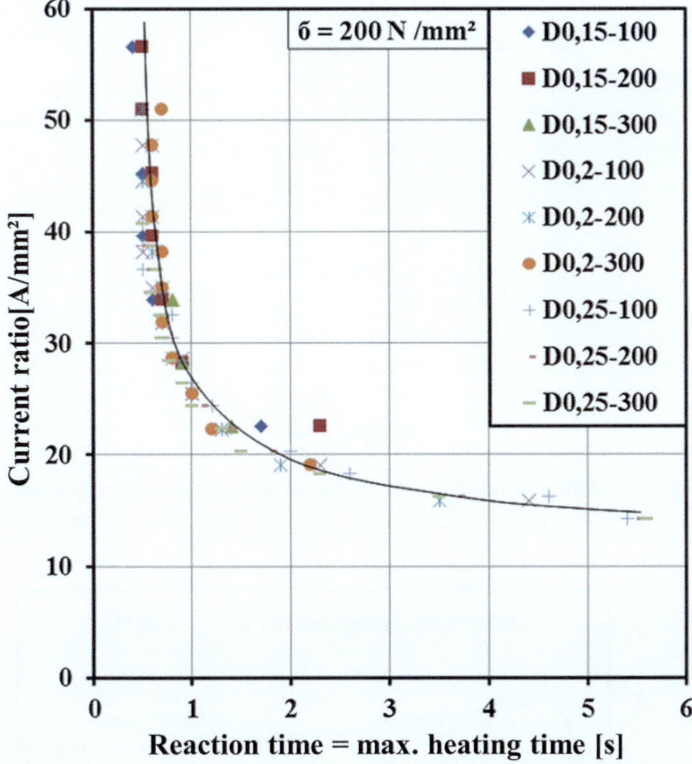

Fig. 4.10 Reaction time of shape memory wires (thickness 0.15–0.25 mm) depending on electric current density [4]

Y-axis indicates heating time, while the cooling for different wire diameters is plotted on the ordinate.

Individual test results are grouped for different wire diameters, so that the influence of the wire diameter W_d is immediately recognizable. For example, a 0.15 mm thick shape-memory wire can be activated by a current of 0.6A in 0.25 s, while 4 A are required to activate a 0.4 mm thick wire in the same time.

As the diameter doubles, the electrical excitation current is quadrupled for the same reaction time. Looking at the cooling period, the independence of this effect from the excitation current is striking. These two effects are therefore decoupled from one another at a controlled current flow without overheating.

The variance of phase transition temperatures is only of minor significance to the controlled electrical activation of shape memory actuators. It does, however, effect the retransformation speed, which is significant when considering system dynamics. To understand the problem, an example at high ambient temperatures (e.g., $T_a = 85$ °C) is given. 85 °C were chosen here because of the temperature specification for actuators in interiors of motor vehicles. In Fig. 4.12, a binary NiTi SMA wire has been measured with an A_F temperature of 90 °C in a stress-free state.

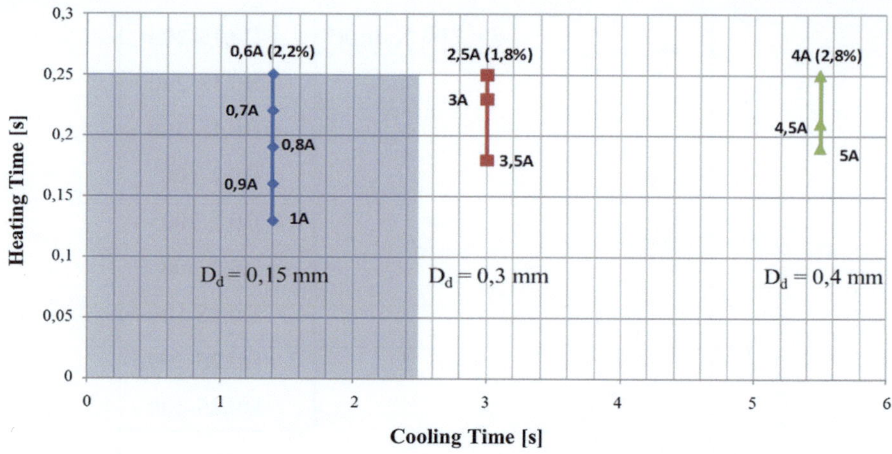

Fig. 4.11 Overview of the dynamic behavior of shape memory wires with different thickness [3]

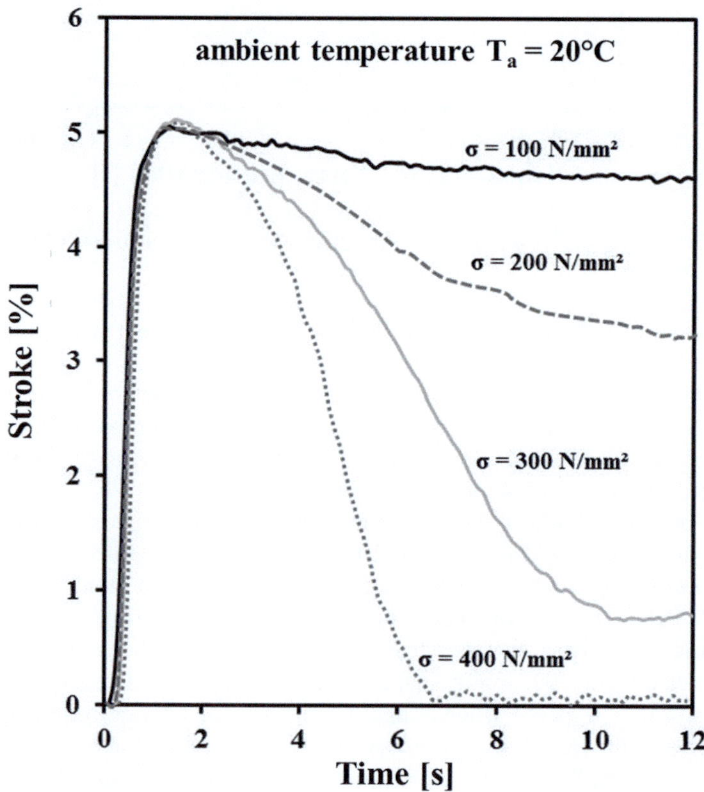

Fig. 4.12 Dynamic actuator behavior under varying mechanical restoring tensions at Ta = 85 °C [1]

Samples ($W_d=0.3$ mm) were electrically controlled by a heating time of 1 s, so that the electrical heating power was automatically switched off after reaching A_F.

The characteristics show a clear dependence between the dynamic system response and the applied mechanical stresses. The special case of $T_a>M_s$ is established at a mechanical recovery stress of 100 N/mm². The shape memory specimen contracted fully with electrical activation, but a retransformation to martensitic state was not possible. Because the M_s temperature of the biased SMA wire is approximately 70 °C, no reverse transformation can start in this case. By increasing the bias load, the conversion temperatures increase by about 1/8 °C (N/mm²). Thus, the 200 and 300 N/mm² pre-stressed shape memory wires retransform only partially. The M_F temperature still lies under the ambient temperature. A full retransformation has been reached at a load of 400 N/mm². This resembles a M_F temperature of about 90 °C. The gradients of retransformation are also dependent on the mechanical bias stress. With increased restoring load, an increase of the retransformation speed can be observed as well.

When considering heating rates, a slight delay of the heating curves with increasing bias loads are striking. The austenite temperatures rise with the bias loads, and therefore the heating process is also delayed.

The following chart (see Fig. 4.13) shows the complementary experiment at increasing ambient temperatures and a constant restoring load of 100 N/mm². Again, a current pulse of a maximum duration of 1 s was chosen, which is automatically switched off after reaching the A_F temperature. While the functionality of a

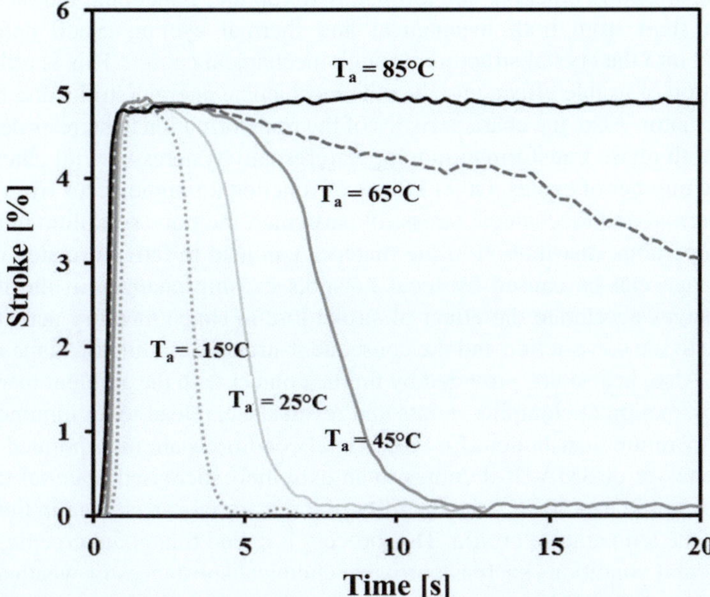

Fig. 4.13 Dynamic actuator behavior under varying ambient temperatures and a constant mechanical load (100 N/mm²) [1]

system with 100 N/mm^2, resetting load of up to 45 °C and ambient temperature is ensured (despite the varying cooling gradient), the retransformation is slowed down from 65 °C in a significant way.

The difference between the M_S temperature and the ambient temperature evens out more slowly, and parts of the sample are also not converted back to martensite. The reason lies in the M_F temperature level, which is below the ambient temperature. At 85 °C the M_S temperature is not even reached, which disables any retransformation. With respect to the heating process, it should be noted that an increased ambient temperature in a constant load results in a faster activation. Since the thermal difference between the ambient temperature and the A_F temperature is naturally lower at high temperatures, the retransformation is finished earlier at constant heating times and current levels. This can cause damages to SMA actuator systems (called overheating). Such fatigue effects will be discussed in the chapter "Fatigue" under SMA actuators.

4.4 Fatigue of Shape Memory Actuators

The fatigue of SMA is a multi-causal phenomenon, namely a combination of structural and functional fatigue [5] details the fatigue behavior on a crystalline level. The causes and effects of functional fatigue are shown in Fig. 4.14.

A structural failure occurs from mechanical overload, which leads to the formation of constrictions, cracking and ultimately to rupture. Functional fatigue, in contrast, can stem from both mechanical and thermal cycling. Such defects are introduced into the crystal structure through mechanical causes. This is reflected in the reduction of usable effect, and thus the mechanical energy (stroke and force) of a SMA actuator. Also, the characteristics of the transformation hysteresis depend on fatigue. Both phase transformation temperatures and hysteresis width change with increasing number of cycles. Other effects of functional fatigue occur from interaction of thermal and mechanical stress. In particular, the polycrystalline inhomogeneous temperature distribution in the material can lead to fatigue-related damage. Such damage can be caused by local hotspots causing changes in the material. These changes accelerate the effect of stroke loss in shape memory actuators and contribute to the constriction and the consequent structural failure of shape memory elements. Also, heat sinks, provided by fluidic contact with the ambient medium, or by heat dissipation at clamping points and reversals, can lead to an inhomogeneity in the temperature distribution. Environmental conditions are incorporated into the fatigue behavior of SMA. If it comes to an extremely increased material temperature (for example $A_F + 50$ °C), the ductility of the material can change in the middle section of the temperature profile. This process is called functional creeping. Other environmental conditions such as corrosive chemical substances or weather conditions have no great influence on SMA, so they may withstand in most applications in industrial engineering. The most important factors influencing the fatigue behavior are summarized in Fig. 4.15.

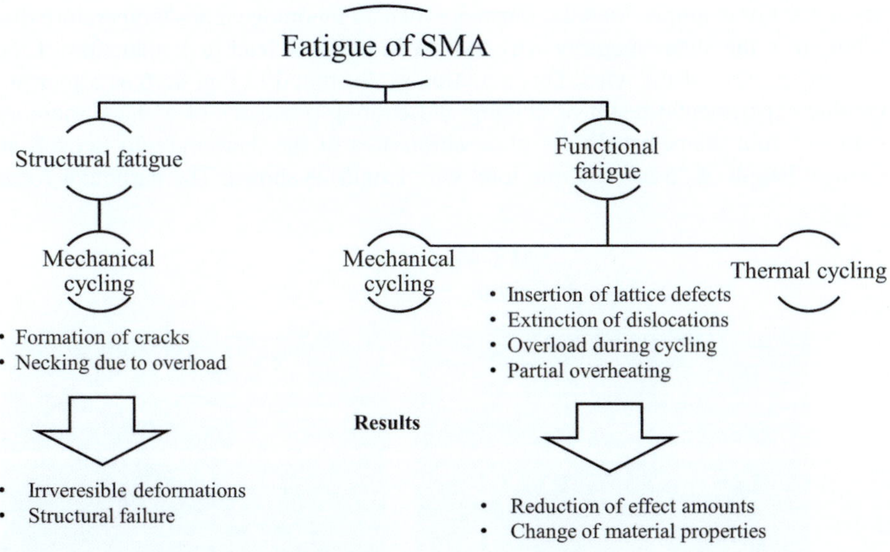

Fig. 4.14 Fatigue of shape memory alloys

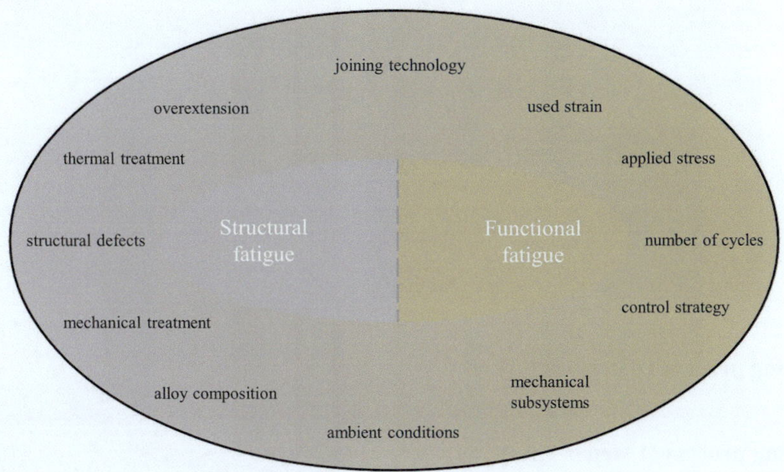

Fig. 4.15 Influences on fatigue behavior of shape memory alloys

4.4.1 Influence of Joining

A large heat sink can be a major cause for inhomogeneous phase transformations. Hence, the connection technology, which is a mechanical and thermal connection to a structural part at the same time, influences the dynamic response of SMA

                                                                 S. Langbein and A. Czechowicz

actuators. For example, metallic ferrules evoke an inhomogeneous temperature distribution in the shape memory wire, which in turn can lead to a reduction of the converting parts of the wire. This situation is illustrated in Fig. 4.16 as a thermographic experimental result of heating and cooling processes of a shape memory wire (0.5 mm diameter). In the observation area of the thermography sensor, an element length of 25 (of 100 mm total wire length) is shown. The particular focus

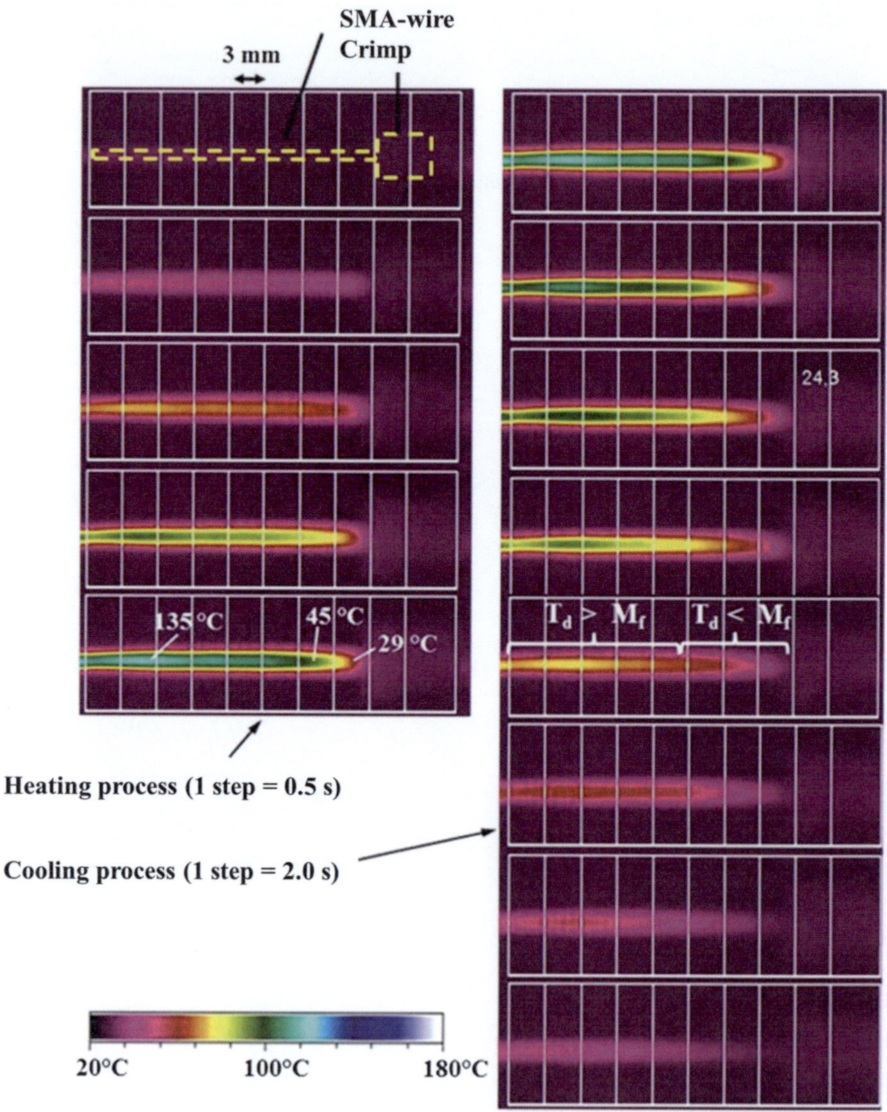

Fig. 4.16 Nonhomogeneous conversion of an SMA sample in close proximity to the mechanical connection through a crimp [6]

lies on the thermal field around the crimp. When the maximal stroke of the actuator is reached (left picture, bottom image), a clear difference between the surface temperatures in the mid-section and the crimp section of the wire can be observed. While the A_F temperature was exceeded in the major part of the SMA wire, the functional element in the area of the crimp remains still in the martensitic state. This inhomogeneity can lead to a mechanical failure. If the SMA wire works against a spring-type load, the critical stress in the system rises with the transformation into the austenitic state of the SMA wire's main part. During this procedure, the increased mechanical load also affects the martensitic parts, which have a significantly lower critical stress level. These parts will be elongated irreversibly, which can lead to cracks and system failure. The inhomogeneity of the retransformation is revealed by thermal analysis. Based on experimental data, shape memory alloy elements cool (at free convection) primarily through thermal conduction into the mechanical connection and not by ambient thermal field (if tested in air at room temperature). This leads to the hypothesis that on one hand the mechanical connection can increase the cooling speed of an SMA, but on the other hand it can also lead to overheating if the electrical system or the electrical activation is not adapted properly. For example, an SMA element has to reach a certain displacement based on calculations. In the first step of electrical adjustment, the required current for a certain heating speed has to be determined through experiments. While increasing the heating current, a system designer will notice a large stroke (main usable stroke) and additional reachable stroke. This additional stroke is less than 10 % of the usable stroke and stems from the transformation of the wire parts in the crimp area.

These parts transform their state while the mid-section of the actuator is already overheated. Therefore, it is necessary to distinguish the maximal stroke and the unavoidable stroke resulting from the extensive overheating. This effect occurs often as the temperature peaks in the central wire section ($T_w > A_F$) and will damage the actuator.

Another thermal fatigue effect is caused by high temperature overheating, which can be reached in the mid-section of the SMA element during attempted heating of all wire elements. During this action, the temperature in the overheated part could reach more than 250 °C, which nearly resembles the heat-treatment temperatures. At these temperatures, the SMA element, which is stressed by an actuator load, becomes more ductile. This also damages the material permanently.

4.4.2 Influence of Stroke and Load

Stroke and pre-stress on the actuator, as well as applied external forces such as interference by a malfunction or changes in load state, affect the fatigue of SMA actuator systems directly. The more the load increases, the more the lifetime of the SMA will decrease. With the same pre-strain, the actuator material contracts completely during the first cycles corresponding to the pre-stretched amount. This reduces the maximum contraction until it settles itself into a seemingly constant level. At high mechanical loads, a structural failure can then also occur from resulting cracks and defects in the material. An increasing mechanical strain is also a risk of introducing irreversible defects in the actuator system, since a larger deformation needs to be retransformed in the martensitic microstructure of austenite by phase transformation [3]. Figure 4.17 shows a simplified overview of the fatigue behavior of shape memory actuators.

The three lifetime areas (N < 30 k, 30 k < N < 100 k, N > 100 k) can be adjusted by a careful selection of mechanical pre-strain and stroke. If one chooses, for example, a first pre-strain of 3.3 %, differing lifetimes and fatigue behavior can be achieved with different mechanical stresses. The experiment in Fig. 4.18 shows that different lifetimes can be achieved at a deformation of 3.3 % at a load of 400 N^2 only 32,000 switching cycles can be reached, while an actuator with the same pre-strain and a pre-stress of 100 N/mm^2 has a substantially longer lifetime. Fatigue-enhancing changes can also occur as a result of wrongly engineered peripheral systems. For example, systems can generate converters or system-friction, which causes stress peaks leading to a rapid increase of the shape memory wire. The same is caused by blockades or overloads applied by peripheral systems, for example through malfunction. Fatigue behavior of NiTi is shown in Fig. 4.18, outlining the lifetime of a ternary NiTiCu, compared to a binary NiTi alloy, which is presented in Fig. 4.19, at comparable stress–strain loads. While the stroke of a NiTi alloy decreases greatly with a rising number of cycles, the NiTiCu alloy seems to inhabit slower fatigue behavior. This is caused by copper, which has a stabilizing effect on the NiTi, with regards to cracks and ductility. Nevertheless, the maximum pre-stress of NiTiCu is significantly lower than that of binary NiTi (about 200–250 N/mm^2).

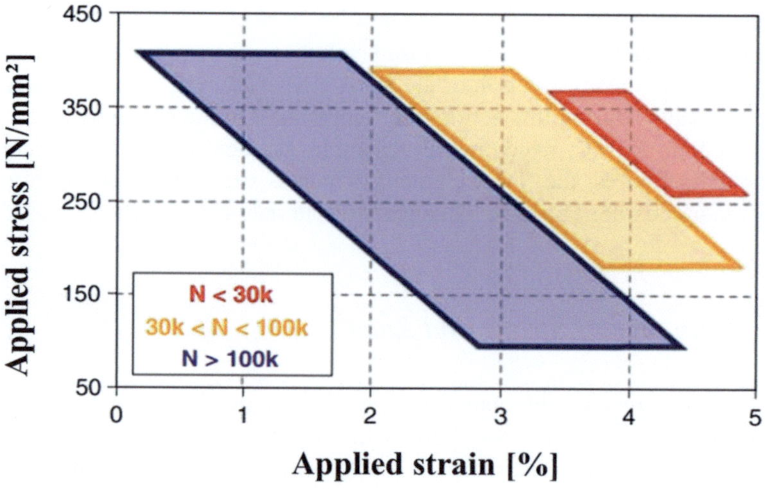

Fig. 4.17 Life chart for SMA lifetime [7]

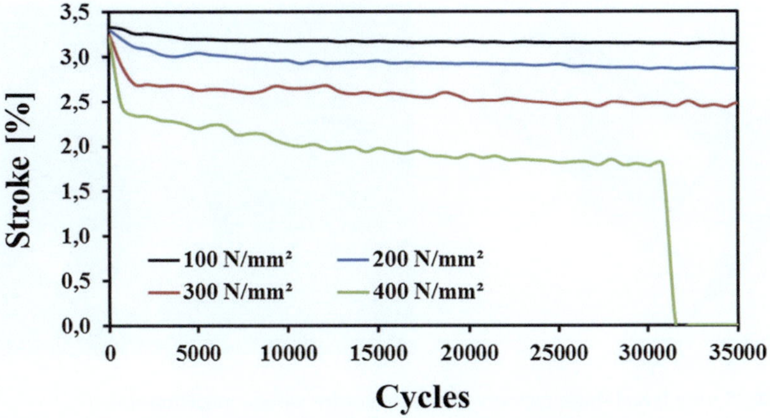

Fig. 4.18 Fatigue testing of NiTi wires with different mechanical stresses [3]

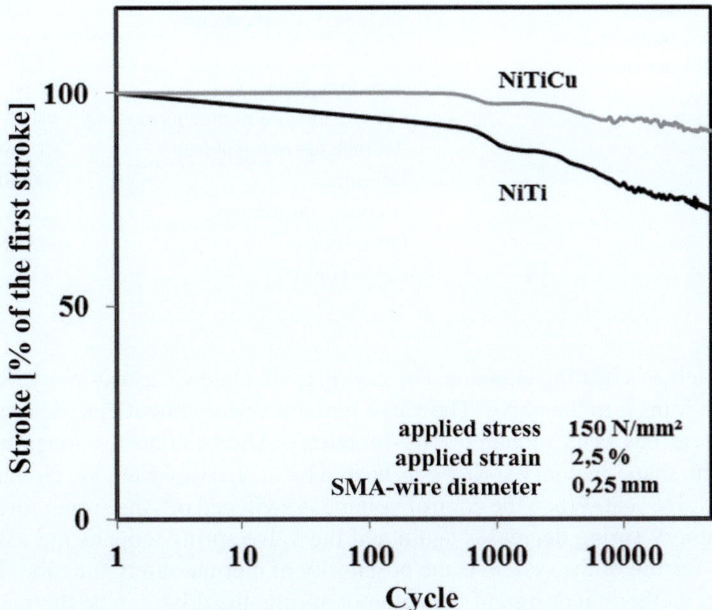

Fig. 4.19 Lifetimes of binary NiTi and ternary NiTiCu at the same conditions [3]

4.5   Designs of SMA Actuators

4.5.1   Spring Actuator with Heating Element

This spring-based actuator (see Fig. 4.20 and Table 4.1) is used as an electro-thermal drive for 2/2-way valves. The SMA drive is externally accessed through electricity and, at its core, consists of a pressure plate with a spring made of a shape memory

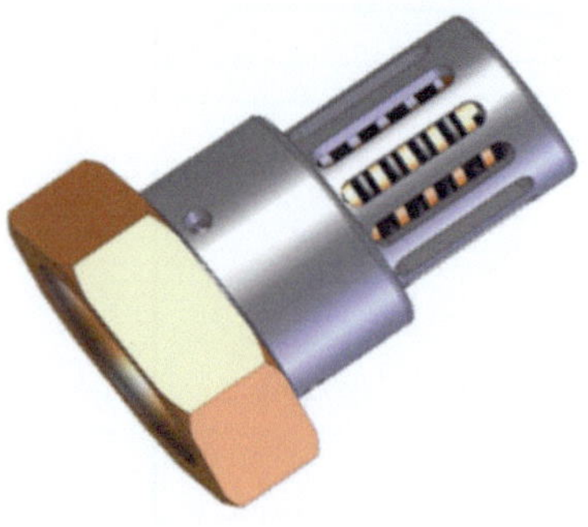

Fig. 4.20 Spring-based shape memory alloy actuator for valve applications [3]

Table 4.1 Technical specifications of the shape memory alloy actuator

Technical specifications	
Stroke	3 mm
Actuating force	60 N
Closing time (at room temperature)	<40 s
Average operating power	<1.5 W
Lifetime	200.000 cycles
Ambient temperature	<50 °C
Dimensions	ø33×50 mm
Connection	M30×1.5

alloy, as well as a heating element. The casing is attached via a customary M30×1.5 screw that joins it to the valve. The valve remains open without electricity after the drive system has been mounted (NO-function). After a control voltage has been applied, the shape-memory spring gets heated by the heating element, elongates and shuts the valve seat. When the control voltage is switched off, the spring force of the shape memory spring decreases again and the valve spring reopens the valve seat. A special feature of the system is the possibility of thermal safety function. Because of the special thermal design of the actuator spring, the drive can be thermally activated due to overheating of the system.

Features

- Spring-based drive for valve systems in panel heaters
- Control by electrical energy supply
- Safety function in overheating of the system
- Small installation space
- Noiseless actuation

4.5.2   Standardized Arc-Shaped Wire Actuator

The wire-based actuator (see Fig. 4.21 and Table 4.2) is controlled externally through electricity and consists of a shape memory wire, a return spring and a correcting element. The shape memory wire has an arcuate shape in the installed condition. During the actuating movement, the angle of the arcuate arrangement change. As a result a conversion of actuating force in displacement is achieved. Depending on its form of application, the correcting element can be executed as a bolt or slide. The casing can be screwed on or adhesively attached. This element is activated by an external power supply, so that the shape memory wire is heated internally by means of the internal resistance heating. Since the wire is activated electrically, it heats up and contracts, which sets the correcting bolt in motion. It returns to its original position upon cooling through the load of the return spring. This process is dependent on the external surroundings.

Features

- Actuator for electrical controlled unlocking systems in motor vehicles (fuel doors, glove compartments, backseats) or in buildings (locking systems, fire-protection appliances)
- Drive for hydraulic or pneumatic valves
- Small installation space
- Noiseless actuation
- Low weight
- High degree of electromagnetic compatibility

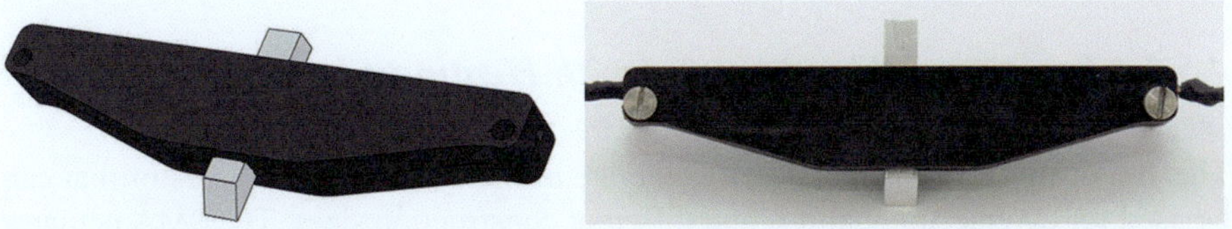

Fig. 4.21 Arc-shaped shape memory alloy actuator for unlocking or valve applications

Table 4.2 Technical specifications of the shape memory alloy actuator

Technical specifications	
Stroke	4 mm
Actuating force	7 N
Closing time (at room temperature)	<5 s
Average operating power	<10 W
Lifetime	100.000 cycles
Ambient temperature	<80 °C
Dimensions	65×25×5 mm
Weight	7 g

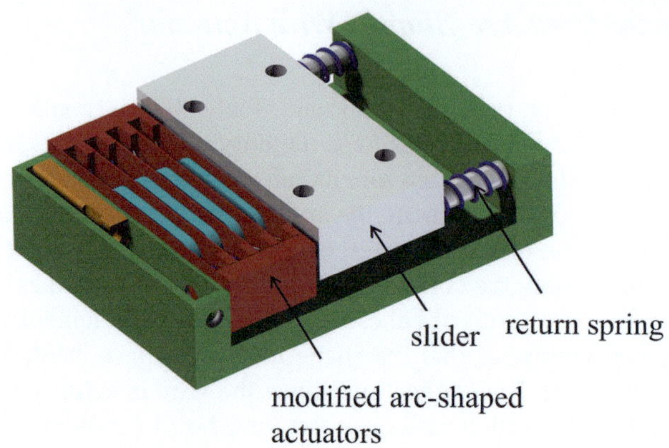

Fig. 4.22 Positioning system consisting of standardized SMA-actuators [8]

An advantage of standardized actuators is the potential to combine them to a modular system. By interconnecting the actuators there is the possibility to increase the stroke and for stepping motion. For the application of a modular adjusting axis (see Fig. 4.22), the arc-shaped actuators are coupled serially to a module. During the heating phase, the actuators press their movable shafts toward the slider. The actuators force is strong enough to move the slider against the return springs. The four standardized actuators can provide four different displacements: 1 mm, 2 mm, 4 mm, and 8 mm. In combination, the module is capable of positioning between 0 mm and 15 mm in 1 mm steps.

4.5.3   Integrated Wire Actuator with Heating Element

The aim of this SMA-actuator is to generate a simply constructed actuator that can be used for diverse applications and different construction sizes. The SMA actuator is made up of the following components which can be configured according to the specifications: SMA -element, plastic carrier, and heating element (dependent on wire strength). Thereby, the plastic carrier structure fulfils different functions within the actuator system. On the one hand, it is used for the mechanical fixing of the SMA wire; on the other hand, it effectuates the reset of the SMA element. The reset forces of the SMA carrier were generated through the spring stiffness of its flexure hinges and were adjusted to the martensite plateau of the SMA wire. Thus, an additional bias spring is not necessary. Another function is the variably controllable conversion of force in displacement [9]. Moreover, the possibility to provide different sizes and load conditions can be achieved by changing the wire diameter and the geometry of the plastic carrier. At low actuating forces a thin SMA element can integrated which can be heated directly. The actuator system is shown in Fig. 4.23 and Table 4.3.

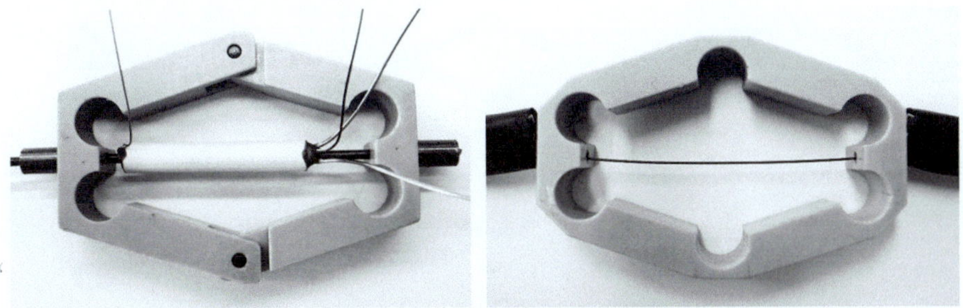

Fig. 4.23 Wire-based shape memory alloy actuator for valve applications

Table 4.3 Technical specifications of the shape memory alloy actuator

Technical specifications	
Stroke	2.5 mm
Actuating force	50 N
Closing time (at room temperature)	<20 s
Average operating power	<3 W
Lifetime	200.000 cycles
Ambient temperature	<50 °C
Dimensions	60×40×5 mm
Weight	10 g

Features

- Wire-based drive for valve systems in panel heaters
- Simple design with a multifunctional plastic carrier
- Control by electrical energy supply
- Small installation space
- Noiseless actuation

In today's version, the connection of the wire is made by crimps. In the future, the integration of the actuator element in the plastic injection molding process will be developed for mass production. Also, the shape of the heating element must be fitted to the demands of such process. In addition, actuator system's variability may still be increased by connecting the carriers in series. For example, it is possible to increase the stroke or to realize a stepped motion. By using injection-molded parts, various realization possibilities are provided. Figure 4.24 shows several concepts.

The drive is designed for use in electrical valves for underfloor heating. The complete valve drive, including electronics, sensors and the housing is shown in Fig. 4.25.

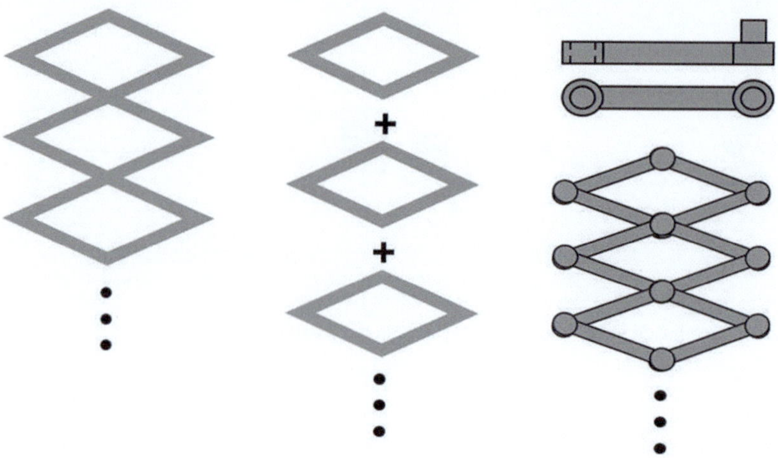

Fig. 4.24 Concepts for the series connection of the plastic carriers [9]

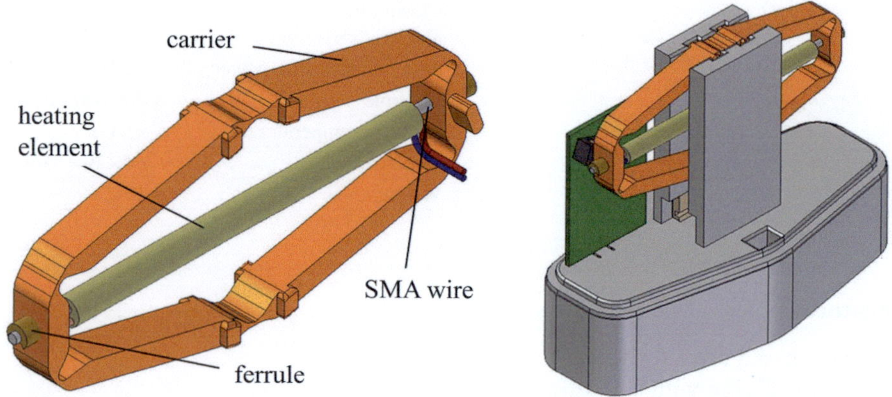

Fig. 4.25 SMA-drive and complete system of the valve drive [9]

4.6 Actuator Systems Compared

4.6.1 Electrical Drive Systems

4.6.1.1 Electric Motors

Electric motors belong to the class of electromechanical transducers. They convert electrical energy into mechanical energy. In electric motors the force is generated by a magnetic field which results from magnetization field in a conductor (wound as a coil) converted into motion; so that the electric motor is the counterpart to the generator which converts motion energy into electrical energy. Today electric motors are used in large numbers in machines, robots, toys, household appliances, electronic equipment, etc. The great importance of the electric motor for today's modern industrial society is also reflected in the energy consumption. Electric motors cause over 50 % of electricity consumption in industrial and household systems. Electric motors are thereby used mostly for rotary motion tasks, but they can also perform translational movements by the usage of transducers like linear worm gear mechanisms. In this context, electric motors are also a competition with shape memory actuators, which are mostly used for their operation for linear movements. Due to the need of translation elements like gears, electro motoric drives are more complex. In addition, transmission causes additional noise and increases the weight of the system. Small engines with a plastic gear may now be offered at low prices. Therefore, it is possible to find these drives in many applications in the automotive industry. These include, for example, unlocking mechanisms for fuel doors, glove compartments, or safety systems. But also the door locking systems are operated by small electric motors.

Particular attention is, however, layed on the ambient temperatures, the target costs and the cyclical dynamics, since the electric motors can be in advantage regarding these factors. Table 4.4 lists the major differences in applications between SMA drives and electric motors. The advantages of SMA drives are highlighted in green, the disadvantages red, and neutral properties in yellow.

4.6.1.2   Solenoids

Solenoids also belong to the class of electromechanical transducers. They usually consist of a hollow coil in which a magnetic field is generated when current flows through the coil. A plunger transmits the actuating movement to the outside. Electromagnets are simpler than electric motors. They generate a translational actuating movement and need therefore no transmission. Electromagnets represent today a standard and are used in many applications for simple positioning tasks. These include, for example, valve actuators. Significant disadvantages of electromagnets can relate to the increased costs compared to the electric motor and the decreasing force on the actuator displacement. The cost of electromagnets also will rise in the future. This is evoked by rising prices of "rare earths" like niobium which are often needed for these actuators.

Similarly to the electric motors, Table 4.5 lists the major differences in applications between SMA drives and solenoids. The advantages of SMA drives are highlighted in green, the disadvantages red, and neutral properties in yellow.

4.6.1.3   Electrified Expansion Elements

This type of expansion elements is unique among electric drives since the electrical energy is converted into thermal energy and finally into mechanical energy. However, the energy conversion of electrically activated SMA actuators resembles

Table 4.4 Major differences between SMA drives and electric motors

	Shape memory actuator	Electric motor
Benefits	Noiseless operation	Operating noise by rapidly rotating gears
	Less installation space in at least 2 dimensions	Larger installation space
	Low weight due to the largest weight-specific working capacity	Higher weight at low actuating forces or small displacements
	Simple construction and small number of components	Complex construction with many components
	Optimum electromagnetic compatibility	Poor electromagnetic compatibility
	Resistant to environmental influences, such as dust and humidity	Resistant to environmental influences, such as dust and humidity, is connected with expenses (sealing)
	Great design flexibility	Restriction of shaping by active principle
	Possibility of integration of the actuator into polymer structures through injection molding	No possibility of integration of the actuator into polymer structures through injection molding
	Possibility of self-monitoring by resistance measurement	Limited possibility of inherent self-monitoring
Neutrally	High and virtually unlimited actuating forces possible	High actuating forces occur due to high gear ratios
	Currentless hold the position by external latching mechanisms	Currentless hold the position by self-locking gears
Handicaps	Low application temperatures for NiTi standard alloys (about up to 353 K)	Higher application temperatures, depending on the materials used (insulating varnish, coil body, potting material)
	Low cyclical dynamics of less than 1 Hz (micro-drives less than 10 Hz)	Higher cyclical dynamics (depending on the gearing structure)
	Small displacement through a limited actuation	Long displacement ranges through endless actuation
	Limited lifetime (at most up to 1 million cycles)	Longer lifetime possible (depending on the mechanical design and the construction of the motor)
	Usually higher priced	Low priced through mass production

these two stages. Electrically activated expansion elements are an attempt to use thermally activated systems as externally controlled electrical actuators. This solution is only a temporary system solution until alternative concepts for smart actuators are established. Additionally, it disposes externally heated expansion elements mainly in heating technology. These include, for example, actuators for valves for

Table 4.5 Major differences between SMA drives and solenoids

	Shape memory actuator	Solenoid
Benefits	High and virtually unlimited actuating forces possible	Limited actuating forces
	Constant release of power	Actuating force is dependent on the displacement
	Noiseless operation	Operating noise by end stops
	Less installation space in at least 2 dimensions	Larger installation space
	Low weight due to the largest weight-specific working capacity	Higher weight at the same actuating force
	Simple construction and small number of components	More complex construction and a higher number of components
	Optimum electromagnetic compatibility	Poor electromagnetic compatibility
	Resistant to environmental influences, such as dust and humidity	Less resistant to environmental influences, such as dust and humidity
	Great design flexibility	Restriction of shaping by active principle
	Possibility of integration of the actuator into polymer structures through injection molding	No possibility of integration of the actuator into polymer structures through injection molding
	Possibility of self-monitoring by resistance measurement	Limited possibility of inherent self-monitoring
Neutrally	Usually high priced	Usually high priced
	Small displacement through a limited actuation	Small displacement through a limited actuation
	Currentless hold the position by external latching mechanisms	currentless hold the position by permanent magnet
Handicaps	Low application temperatures for NiTi standard alloys (about up to 353 K)	Higher application temperatures (about up to 473 K)
	Low cyclical dynamics of less than 1 Hz (micro-drives less than 10 Hz)	Higher cyclical dynamics (up to 1 kHz)
	Limited lifetime (at most up to 1 million cycles)	Longer lifetime possible (up to 5 million cycles)

underfloor heating. Heated expansion elements have the advantage of high actuating forces due to old standards. Disadvantages are the very sluggish actuation and the need of a mounted heating device. Often heating elements are only glued to the expansion element with a relative high potential for errors. A substitution of these elements by SMA actuators therefore seems sensible. However, still hindrances against the assertion of SMA in the field of heating systems exist. In comparison to solenoids and electro motors, the advantages of SMA to electrified expansion elements focus more dynamic aspects.

4.6.2 Thermal Actuators

4.6.2.1 Thermo-Bimetals

Thermo-Bimetals use the effect of thermal expansion to convert thermal energy into mechanical energy. Thus the operation is very similar to that of SMA. Thermo-bimetals are layer composites which consist of at least two components with different thermal expansion coefficients. As the one component expands more upon heating than the other, the composite generates a temperature-dependent deformation of the thermo-bimetal. Thermo-bimetal components are designated by the smaller thermal expansion as a passive component and the component having the larger thermal expansion as the active component. Thermal bimetallic strips have a good linearity range. The linearity is more than ±5 % different from the deflection, which is calculated from the nominal value of the specific thermal curvature and the nominal thickness. Outside this range these elements also exhibit a nonlinear behavior. The operation of bimetallic composites and possible combinations are shown in Fig. 4.26.

If the main features of SMA actuators and thermal bimetallic are compared, it can be stated that SMA elements should be preferred in applications with a high actuating forces or by the necessity of an electrification or an integration of sensor and monitoring functions. Before utilization system costs, lifetime and thermal requirements have to be compared. In these categories thermo-bimetals are beneficial. Table 4.6 sums up the major differences of SMA to thermo-bimetals.

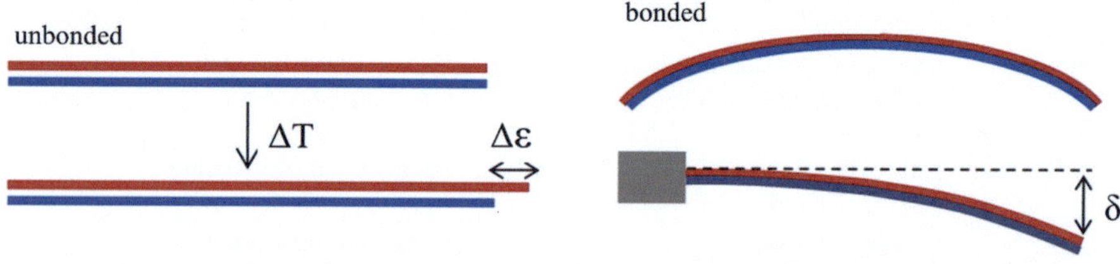

active component	passive component	k 10⁻⁶ [K]	linearity range [°C]	operating limits [°C]
FeNi20Mn6	FeNi36	28,5	-20…+200	450
MnCu18Ni10	FeNi36	39,0	-20…+200	350
MnNi15Cu10	FeNi32Co6	42,6	+20…+230	350
X12CrNi18.8	FeNi32Co14Ti1,5	18,0	-20…+400	650
X12CrNi18.8	X8Cr17	9,5	-20…+600	550

Fig. 4.26 Fundamental designs and data of thermo-bimetals [3]

Table 4.6 Major differences between SMA and thermo-bimetals

	Shape memory element	Thermo-bimetal element
Benefits	High and virtually unlimited actuating forces possible	Limited actuating forces
	Actuation in a small temperature range, easy adjustable through a heat treatment	Actuation in a large temperature range
	Low weight due to the largest weight-specific working capacity	Higher weight at the same actuating force
	Great design flexibility	Restriction of shaping by multi-layer design
	A variety of deformation possibility (tensile, compressive, torsional or bending deformation)	Only the possibility of bending deformation
	Possibility of heating by the inherent resistance	Usually no possibility of heating by the inherent resistance
	Possibility of self-monitoring by resistance measurement	No possibility of inherent self-monitoring
Neutrally	Noiseless operation	Noiseless operation
	Simple construction	Simple construction
	Small displacement through a limited actuation	Small displacement through a limited actuation
	Small installation space in at least 2 dimensions	Small installation space at low actuating forces
	Resistant to environmental influences, such as dust and humidity	Resistant to environmental influences, such as dust and humidity
	Possibility of integration of the element into polymer structures through injection molding	Possibility of integration of the element into polymer structures through injection molding
Handicaps	Low application temperatures for NiTi standard alloys (about up to 353 K)	Higher application temperatures (typically up to 623 K)
	Large hysteresis (approximately 15 K), but with R-phase-transformation only 1–2 K	Nearly no hysteresis
	Limited lifetime (at most up to 1 million cycles)	Longer lifetime possible (typically up to 10 million cycles)
	Resetting element required	Because of the bimetal stress no resetting element required
	High priced	Low priced

4.6.2.2 Expansion Elements

Expansion elements also convert thermal energy into mechanical energy, taking advantage of the effect of thermal expansion. In contrast to thermostatic bimetals, special fluids and waxes are used in these elements. When heated, these fluidic elements' volume increases. The increase in volume is used to generate a stroke.

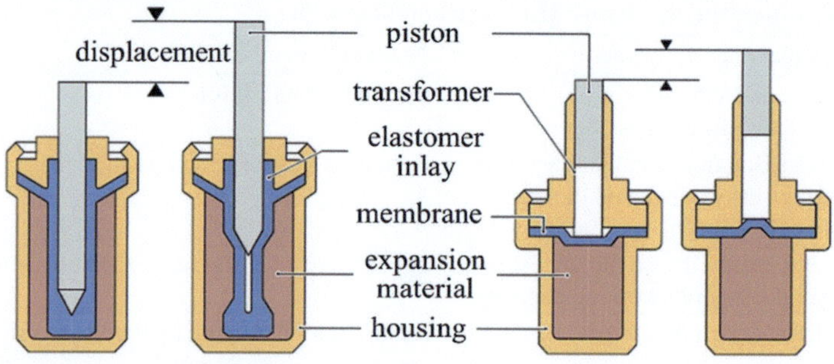

displacement-temperature curve:

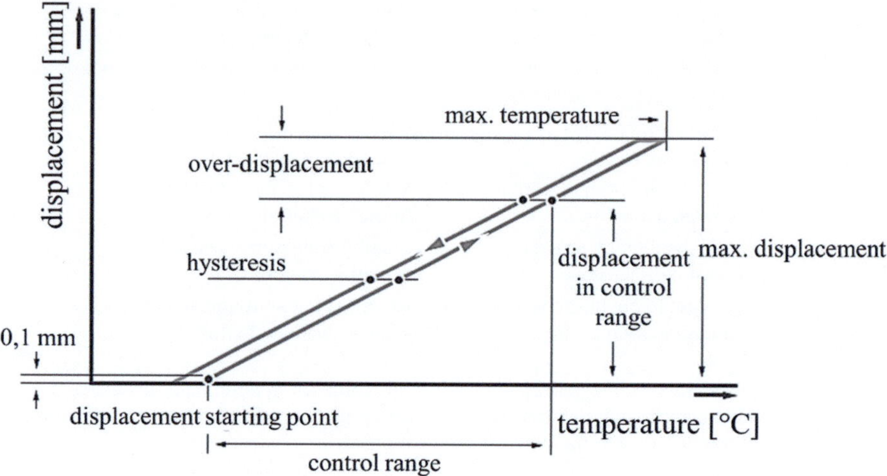

Fig. 4.27 Fundamental design of thermal expansion elements [3]

Expansion elements have a linear stroke–temperature characteristic with a low hysteresis of a few Kelvin. The structure of expansion elements is described in more detail in Fig. 4.27.

For all applications of thermal expansion elements, the selection of the expansion fluid is the crucial parameter. While liquids or solids were used in the past, today special olefins (waxes) have prevailed in the fluidic control area. Wax shows the most relevant temperature range 253–293 K and a very good strain behavior. Low costs and simple processing have contributed to the usage of these elements in most industrial thermal actuating applications at moderate to high forces.

In comparison to thermal expansion elements, SMA are preferable in applications with a small installation space, a high dynamic or by the necessity of an electrification

Table 4.7 Major differences between SMA and expansion or wax elements

	Shape memory element	Expansion element
Benefits	Rapid response to changes in temperature	Slower response to changes in temperature
	High and virtually unlimited actuating forces possible	High but limited actuating forces
	Simple construction	More complex construction
	Low weight due to the largest weight-specific working capacity	Higher weight at the same actuating force
	Great design flexibility	Restriction of shaping by fluid storage
	A variety of deformation possibility (tensile, compressive, torsional or bending deformation)	Only the possibility of deformation by compression
	Possibility of heating by the inherent resistance	No possibility of heating by the inherent resistance
	Possibility of self-monitoring by resistance measurement	No possibility of inherent self-monitoring
	Possibility of integration of the actuator into polymer structures through injection molding	No possibility of integration of the actuator into polymer structures through injection molding
Neutrally	Noiseless operation	Noiseless operation
	Actuation in a small temperature range, adjustable through a heat treatment	Actuation in a small temperature range, adjustable through the medium filling
	Small displacement through a limited actuation	Small displacement through a limited actuation
	Small installation space in at least 2 dimensions	Small installation space at high actuating forces
	Resistant to environmental influences, such as dust and humidity	Resistant to environmental influences, such as dust and humidity
	Large hysteresis (approximately 15 K), but with R-phase-transformation only 1–2 K	Small hysteresis (approximately 1 K)
Handicaps	Low application temperatures for NiTi standard alloys (about up to 353 K)	Higher application temperatures (typically up to 453 K)
	Limited lifetime (at most up to 1 million cycles)	Longer lifetime possible (above 1 million cycles)
	Usually higher priced	Low priced through mass production

or an integration of sensor and monitoring functions. Before utilization system costs, lifetime and thermal requirements have to be compared also by regarding SMA elements as substitutes for thermal expansion elements. Table 4.7 sums up the major differences of SMA to thermal expansion elements.

References

1. A. Czechowicz, *Adaptive und adaptronische Optimierungen von Formgedächtnisaktor-systemen für Anwendungen im Automobil* (Shaker Verlag, Aachen, 2012)
2. A. Czechowicz, J. Boettcher, S. Mojrzisch, S. Langbein, *High speed shape memory alloy activation*. Proceedings of SMASIS conference, Stone Mountain (Georgia), USA 2012
3. S. Langbein, A. Czechowicz, *Konstruktionspraxis Formgedächtnistechnik* (Springer Vieweg, Mannheim, 2013). 3834819573
4. H. Meier et al. Smart control systems for smart materials. J. Mater. Eng. Perform. 20 (2010), 559–563, Springer-Verlag, New York 2011
5. G. Eggeler, E. Hornbogen, A. Yawny, A. Heckmann, M. Wagner, *Structural and functional fatigue of NiTi shape memory alloys*. Mater. Sci. Eng. A, **378**, Elsevier 2004, S. 24–33
6. H. Meier, A. Czechowicz, *Computer-Aided Development of Shape Memory Alloy Actuators as an Approach Towards a Standardized Developing Method*. ASME Conference on Smart Materials, Adaptive Structures and Intelligent Systems (SMASIS), Scottsdale, Arizona, USA 2011.
7. A. Coda et al. *SmartFlex NiTi Wires for Shape Memory Actuators*, Journal of Materials Engineering and Performance, **18**, 691–695 (Springer Verlag, New York, 2009)
8. H. Meier, A. Czechowicz, S. Langbein, *Concepts for Standardized Shape Memory Actuators for Positioning Applications*. Proceedings of the International ASME Conference for Smart Materials, Adaptive Structures and Intelligent Systems (SMASIS), September 2010.
9. S. Langbein, E. G. Welp, J. Sohn: *Development of a Variable and Integratively Structured SMA-Actuator for Multifunktionale Applications*. Proceedings of the 11th International Conference on New Actuators (ACTUATOR), Bremen 2008, p. 402–405

Chapter 5
Sensing Properties of SMA Actuators and Sensorless Control

Stefan Seelecke

5.1 Introduction

In addition to suggesting an application potential for valve actuation, the material behavior outlined in previous sections also motivates further use of shape memory alloys for sensing purposes. This feature is based on the measurement of changes in the electrical resistance, which accompany the contraction and elongation of SMA actuator wires. Of particular interest is the simultaneous application of both effects, leading to so-called "self-sensing" or "sensorless" concepts for SMA actuators. Implementation of such concepts would allow designing valve systems which could be controlled without any additional external sensors, further cutting cost, size, and part count.

The following section first presents basic experimental results to discuss the background of the effect. The applicability to position sensing is then motivated in the context of a novel power control concept, which also includes an electronics platform for its implementation. Further use for flow rate sensing is subsequently motivated, which is of particular interest in the current context of valve applications. The section concludes with a brief illustration of a control scheme based on self-sensing.

5.2 Material Behavior

In order to illustrate the various actuation and sensing properties of shape memory alloys, a series of experiments are discussed below. Figure 5.1 shows the results from temperature-driven experiments using a CTS climate chamber. For these and

S. Seelecke (✉)
Saarland University, Department of Mechatronics Engineering Campus,
66123 Saarbrücken, Germany
e-mail: seelecke@mx.uni-saarland.de

© Springer International Publishing Switzerland 2015
A. Czechowicz, S. Langbein (eds.), *Shape Memory Alloy Valves*,
DOI 10.1007/978-3-319-19081-5_5

                                                                S. Seelecke

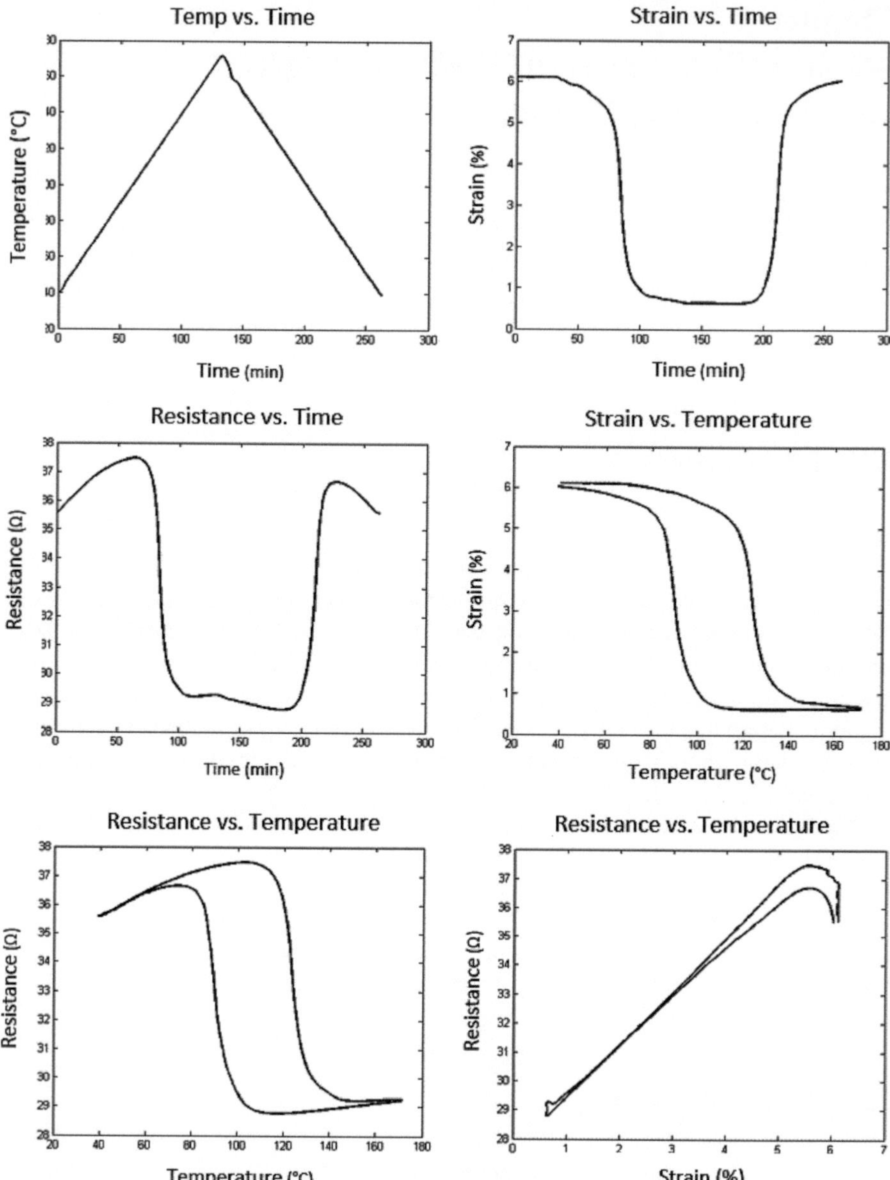

Fig. 5.1 Experimental results of SMA wire resistance measurement

throughout the following experiments a typical, commercially available NiTi wire has
been used for the experiments, Flexinol HT from Dynalloy, Inc. These actuators are
available with diameters ranging from 25 to 500 μm and are cited by the manufacturer
with an austenite start temperature of 90 °C under a mechanical stress of 172 MPa.

The data in Fig. 5.1 were recorded with a 75 μm wire under a constant stress of 300 MPa, the first row showing the temperature input as a linear ramp with a rate of 1 °C/min from 40 to 170 °C along with the resulting time-resolved strain and electrical resistance responses. It should be mentioned here that in the interest of a long and fatigue-free actuator life, such high temperatures and stresses should be avoided in actual operation; they were used here for a clearer demonstration of the underlying effects only. Depending on stress level and alloy compositions, the resistance behavior can become considerably more complex due to the presence of an additional phase called the R-phase because of its rhombohedral crystal lattice structure. For a detailed study of the related phenomena, the reader is referred to [1].

The left and center figure in the second row show the corresponding strain/temperature and resistance/temperature curves. The key observation is the clear hysteretic behavior, which renders the use of NiTi wires as temperature sensors rather complicated as there exists no one-to-one correspondence between actuator stroke and resistance on the one hand and temperature on the other. Fortunately, it can be seen from the right figure that elimination of the temperature results in a dramatically reduced hysteresis between resistance and stroke. In fact, during large portions of the stroke, a nearly linear relationship can be observed; this motivates the use of resistance data as a position signal and provides the basis for the so-called self-sensing effect in NiTi actuators.

5.3 Sensor/Actuator Behavior

For actuation purposes, it is important to not only consider an individual actuator wire but rather a system, which in the simplest case complements the actuator wire by a spring for restoring purposes. Figure 5.2 shows a sketch of such a basic system in the top row and a sketch of an instrumented set up with load cell, displacement sensor, and linear actuator in the center row. A steel cantilever with moderate stiffness is used as a linear spring in this case, permitting simple deflection measurements of its backside, while the linear actuator enables precise adjustment of the wire pre-stress. Specifically, a laser sensor was used together with a linear push–pull actuator. Data were taken with a 50 μm Flexinol HT wire similar to the one used in the previous section.

In the previous section, it also became clear that it is problematic to establish a useful sensor relation between resistance and temperature due to the presence of a hysteresis. The problem runs still deeper if one considers that even the measurement of the temperature itself of a micron-sized wire is no trivial task. Contact methods such as PT-1000 sensors must be ruled out because they would (a) not be able to capture a reliable transient signal because of the minimal contact surface to the wire and their considerable thermal mass and (b) would also impact the wire temperature and hence the actuator behavior. Even though it could in principal be done with contactless thermal imagery methods, this would require high-end equipment and does not represent a practically feasible method either.

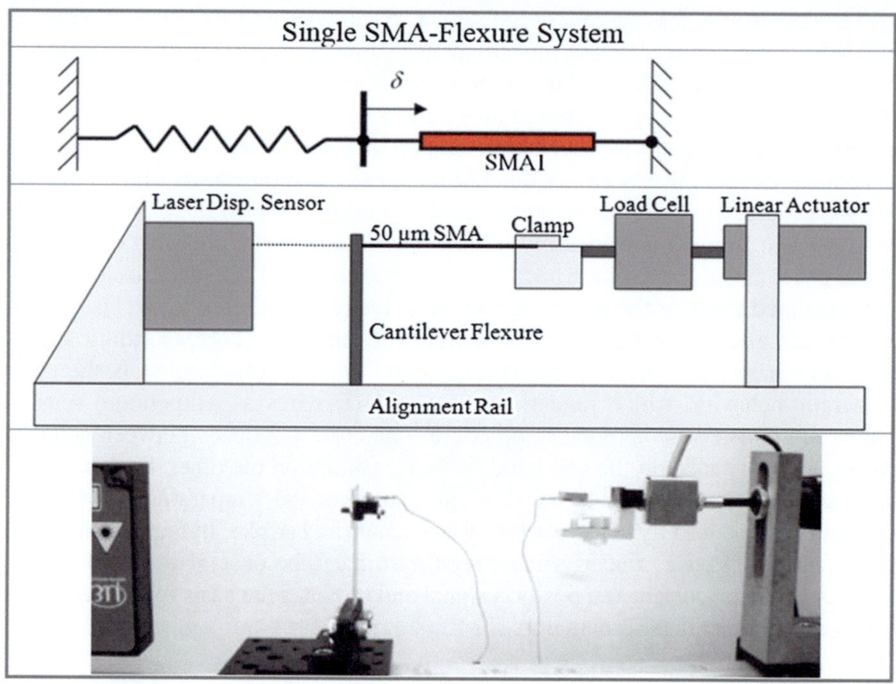

Fig. 5.2 Experimental setup for SMA wire resistance testing

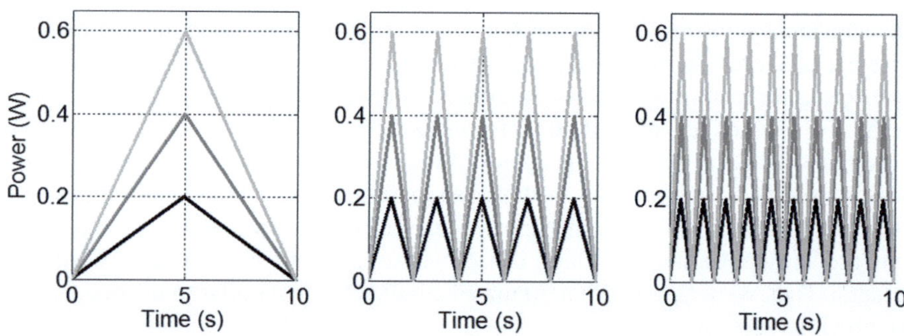

Fig. 5.3 Input power signals shown over 10 s at 0.1 Hz (*left*), 0.5 Hz (*middle*), and 1.0 Hz (*right*)

Additionally, the mechatronics engineer prefers to have a directly controllable quantity at his disposal, which motivates a concept of electric power instead of temperature. In the sequel, the implications of this concept on the actuation and sensing behavior will be illustrated using the set up from above.

To drive the experiments, the wire was electrically heated by multi-linear power signals at various frequencies and amplitudes as displayed in Fig. 5.3.

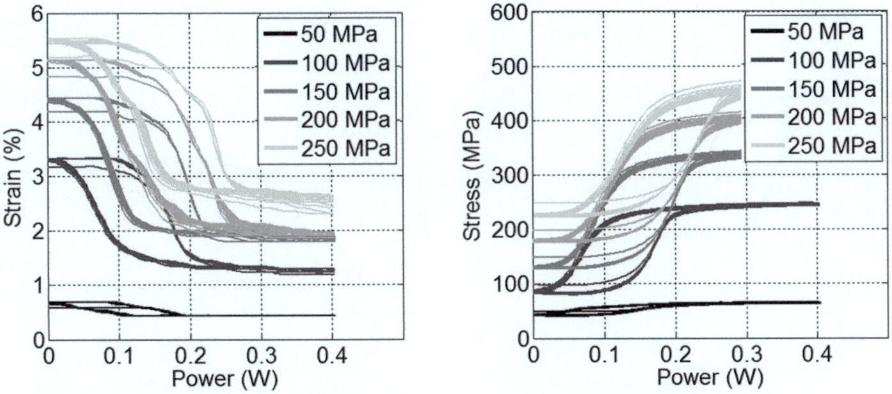

Fig. 5.4 Strain vs. power (*left*) and stress vs. power (*right*) at different pre-stresses

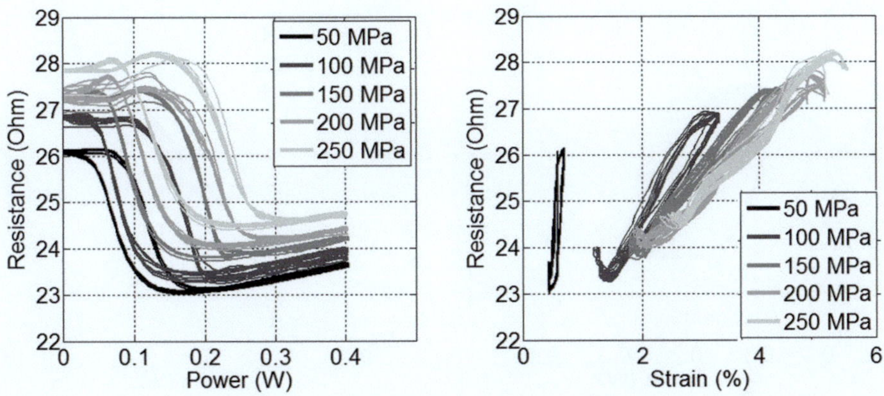

Fig. 5.5 Resistance vs. power (*left*) and resistance vs. stress (*right*) at different pre-stresses

Figures 5.4 and 5.5 show the strain, stress, and resistance responses plotted vs. the electric power for a slow experiment performed at 0.1 Hz at various mechanical pre-stresses. These results resemble the temperature curves that are typically used to characterize the material behavior, and one also obtains the linear sensor relationship again upon elimination of power from the picture.

However, the introduction of the power concept comes at a price. Obviously, power is not a state quantity as temperature is; consequently, there is not a one-to-one correspondence between the two. As a matter of fact, if a constant power signal is applied to the wire, temperature will continuously change with time adjusting to the power input until it reaches a stationary value. For a given heat exchange environment, there is only a correspondence between power and the stationary temperature, which is why we observe the familiar hysteresis curves only for a very slow experiment. Here, the temperature has enough time to always settle to its stationary value.

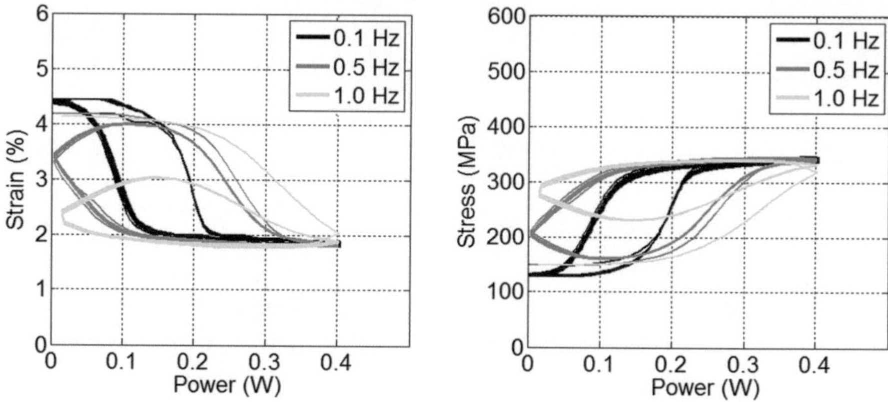

Fig. 5.6 Strain vs. power (*left*) and stress vs. power (*right*) at different frequencies

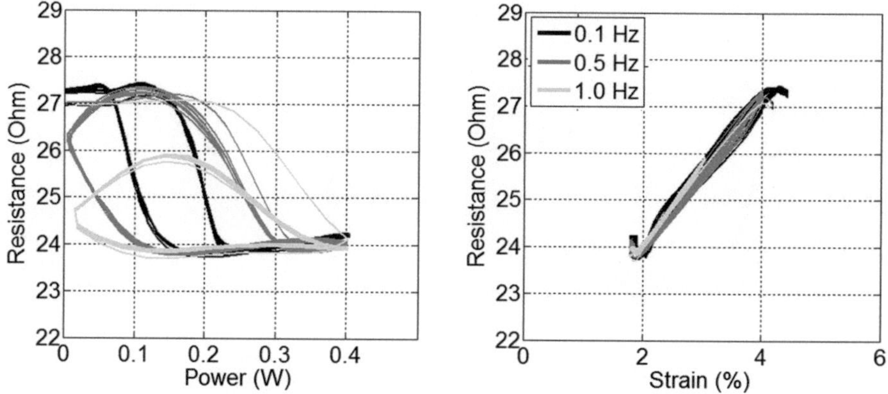

Fig. 5.7 Resistance vs. power (*left*) and resistance vs. stress (*right*) at different frequencies

For higher frequencies, one expects a deviation from this behavior, and Fig. 5.6 displays this deviation for experiments performed at five and ten times the original frequency. With increasing frequency, one observes a widening of the hysteresis because the underlying temperature does no longer have time to reach its stationary value and consequently lags behind the strain and stress values achieved at the same power level at lower rates.

A similar phenomenon can be observed for the resistance behavior, and one might wonder if this will render the self-sensing behavior invalid. Fortunately, however, the resistance-strain sensing property demonstrated for the temperature-driven experiments is not affected by the introduction of the power concept, see Fig. 5.7. Further details and simulations of the experiments presented here can be found in [2].

In addition to the strain-sensing property demonstrated above, there is yet another sensing feature implied by the electro-thermo-mechanical behavior of NiTi wires.

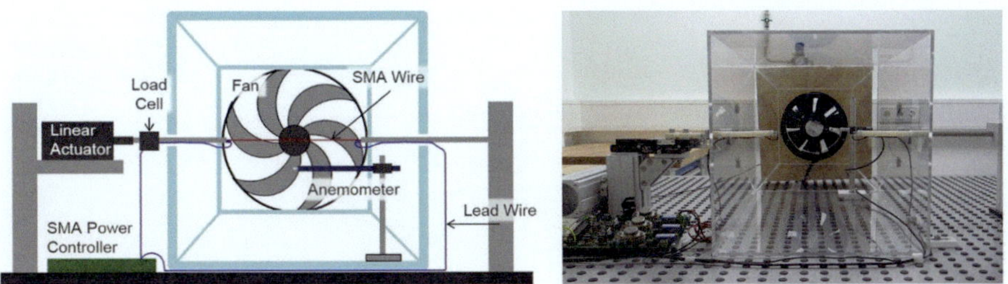

Fig. 5.8 Experimental setup for determination of flow rate of transportation medium

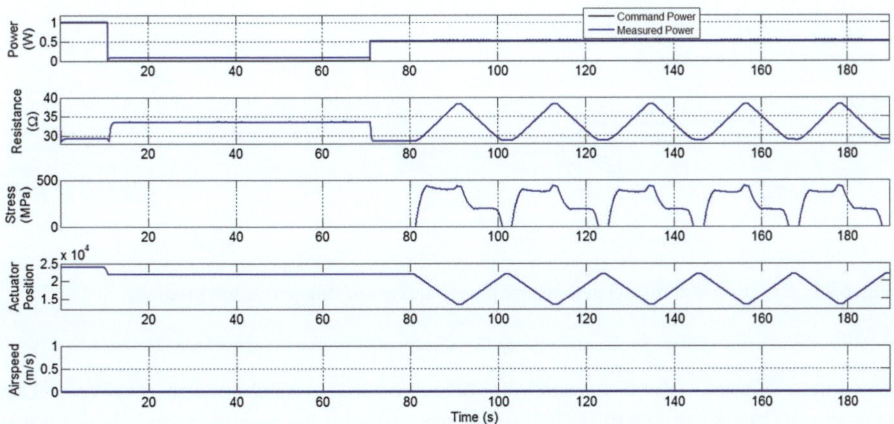

Fig. 5.9 Experimental result for determination of flow rate of transportation medium

For the operation of a valve system, it is oftentimes desirable to measure the flow rate of the transport medium. It can be shown that a NiTi wire is not only capable of actuation and strain self-sensing, but that it can also determine this flow rate, making NiTi a truly multifunctional material.

For the demonstration of this effect, the set up in Fig. 5.8 is used. It consist of a linear motor with position encoder, a load cell, a custom-built power controller, which will be further described in the following section, a wind tunnel with flow straightener and a NI DAQ system.

The set up enables tensile tests under various electric power levels and wind speeds, with a typical experiment for a 75 μm HT Flexinol wire being shown in Fig. 5.9, see [3] for details.

Figure 5.10 shows stress–strain curves and resistance–strain curves at power levels from 0 to 1.0 W under still ambient air condition at 21 °C. The data on the left show the transition from loading and unloading on the tensile martensite branch at low power levels to the full superelastic behavior at intermediate levels and purely austenitic loading/unloading at 1 W. The figure on the right displays the corresponding

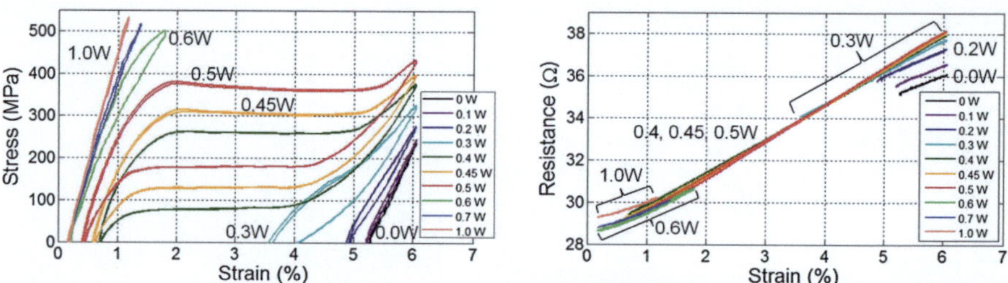

Fig. 5.10 Experimental results of power-dependent SMA characteristics

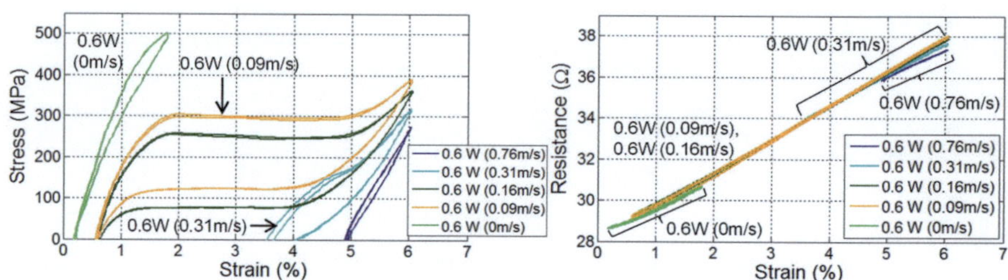

Fig. 5.11 Experimental results at constant power and at varying activation gradients

resistance curves, which reveals the effect of three mechanisms involved. The 0 W-curve (black) shows the effect of deformation on the resistance, while the jump to the 0.1 W-curve (magenta) determines the temperature-dependent part of the resistance change. The intermediate levels finally show the large resistance change due to the phase transformation from martensite to austenite upon further heating. The two key observations here are the magnitude of the resistance change during the phase transformation of roughly 25 % of its nominal value and the fact that even though the loading/unloading curves to the left are characterized by the large super-elastic hysteresis, such hysteretic effects are completely absent in the resistance behavior. The latter effect again motivates the use of NiTi wires as actuators, which can sense their length change without additional external sensor, while the former effect suggest a potential application as a large-deformation strain gauge.

The diagrams in Fig. 5.11 show results from identical experiments, only this time taken at fixed electric power and varying wind speeds, measured by an anemometer. The different wind speeds were chosen such that the stress–strain curves match the power levels from the experiment at zero wind speed, and by identification of these curves the excess power needed to produce them can be related to the corresponding wind speeds. Figure 5.12 shows the results for two different power levels, and it can be seen that both cases give rise to a common fit such that the excess power values can be used to uniquely identify the wind speeds. As a consequence, for a given valve geometry, this effect can be used to determine the flow rate if the actuator was to be exposed to the medium.

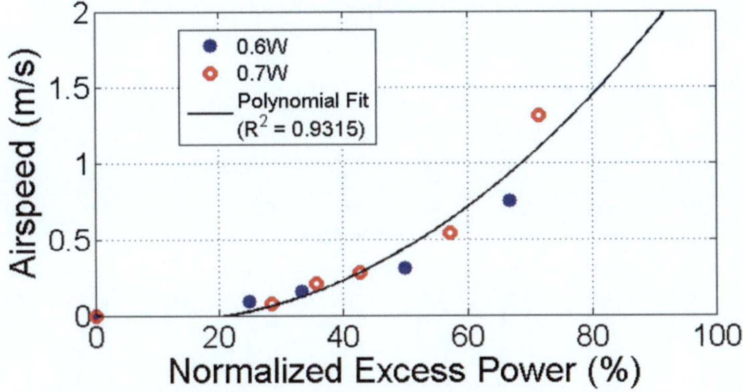

Fig. 5.12 Measured airspeed vs. normalized excess power

Fig. 5.13 Design of SMA control systems with pulse-wide modulation

5.4 Electronics

In the previous section, reference was already made to the electronics used in the context of the presented experiments. Apparently, the driving force behind the actuation of a NiTi wire is a temperature change effected by the Joule heating of an electric current. This temperature change is determined by the applied electric power; however, typical amplifiers provide either current or voltage control modes only. In the case of a constant resistor, this also provides for power control, but for NiTi wires with their complex nonlinear temperature-dependence, this is clearly not the case. For this reason, a special electronic circuit was developed [4], which is shown in Fig. 5.13. Power control is effected by a pulse-width-modulated constant current source, of which the duty cycle is adjusted based on continuous resistance

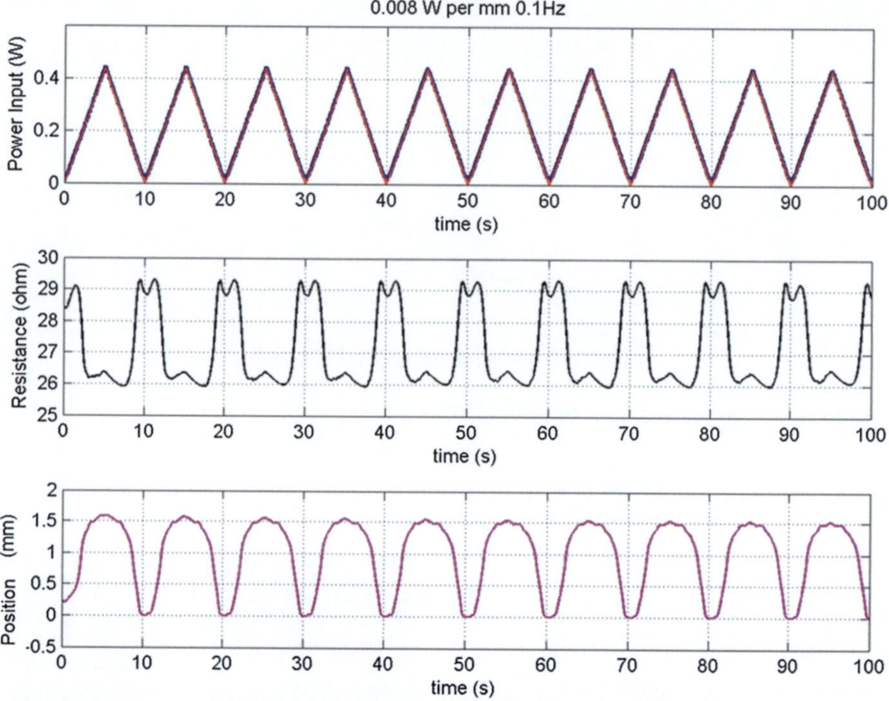

Fig. 5.14 Experimental results of resistance-based control of SMA actuation position

measurements. This mechanism is implemented on a micro-controller and does not only allow for an on-line power adjustment, but simultaneously provides a sensor reading through the resistance measurement, which can be used to determine the actuator position.

Figure 5.14 shows the device outputting a repeated multi-linear power signal (top), demonstrating that the actual signal (blue) is able to closely follow the prescribed set point (red). The center row displays a low-noise, reproducible resistance signal, displaying a typical "up-down-up" sequence in response to a power increase. This is the result of heated martensite with a resistance increase typical for a metal conductor, followed by a drop due to the transformation to the lower-resistance austenite, and a subsequent increase now due to heated austenite. During the cooling phase, this behavior is reversed, and a "down-up-down" sequence completes the cycle. The bottom row shows the stroke of the actuator wire measured by an external laser sensor.

5.5 Control

This last section shows the implementation of a simple PI control scheme to illustrate the feasibility of the self-sensing concept. It uses the SMA wire/flexure spring set up from Fig. 5.2, and it is based on a number of steps motivated by the experimental data presented in the previous sections.

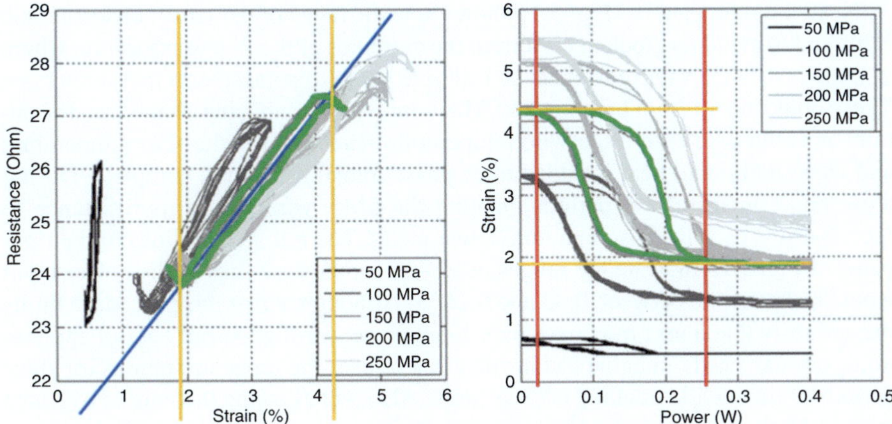

Fig. 5.15 Example of using characteristics of a SMA-spring system to generate a displacement sensor mapping

5.5.1 Resistance to Deflection Sensor Mapping

The first step is to relate the strain change in the wire to a displacement in order to determine the actuator position. The simplifying assumption behind the scheme for mapping resistance to deflection is that resistance relates approximately linearly and non-hysteretically to SMA wire strain while phase transformation is occurring. Under this assumption, the plots in Fig. 5.15 are used to motivate a general scheme for creating a position feedback sensor out of a SMA-spring system. In the plots in Fig. 5.15, the 150 MPa pre-stress case is isolated and used to generate a resistance to displacement mapping. If another pre-stress or pre-strain were used, a separate mapping would be required. The resistance vs. strain plot in the left panel of Fig. 5.15 is qualitatively equivalent to resistance vs. displacement because in the flexure setup shown in Fig. 5.2 strain is calculated directly from the measurement of flexure displacement scaled by the length of the SMA wire. The characteristic for the 150 MPa case is highlighted (green), and a linear curve fit (blue) is generated to approximate the mapping.

The objective of the mapping is to get a function for displacement as a function of resistance. First, in Eq. (5.1) resistance is mapped linearly to strain. Then the deflection of the flexure, δ, is scaled with the change in length of the SMA wire, as shown in Eq. (5.2). Finally, Eq. (5.1) is substituted into Eq. (5.2) and constant coefficients are combined such that $k = L_{0,\mathrm{SMA}} k_1$ and $\delta_0 = L_{0,\mathrm{SMA}} \left(\varepsilon_0 - \varepsilon_{\mathrm{pre}} \right)$ to create Eq. (5.3) that maps resistance directly to deflection.

$$\varepsilon = k_1 R + \varepsilon_0 \tag{5.1}$$

$$\delta = L_{0,\mathrm{SMA}} \left(\varepsilon - \varepsilon_{\mathrm{pre}} \right) \tag{5.2}$$

$$\delta = kR + \delta_0 \tag{5.3}$$

Next, the right panel of Fig. 5.15 shows how the range of the sensor characteristic is limited to avoid the hooked regions at the extremes of the R–ε relationship, where a single resistance value correlates to multiple strain (displacement) measurements. These hooks result from heating the SMA wire without inducing phase transformation, as is the case before the wire temperature reaches austenite start temperature and when additional heat is added after phase transformation is complete. During these non-transformation periods, heating the SMA wire causes its resistance to increase, just as is the case in a non-active metal. The red vertical lines in the right panel of Fig. 5.15 indicate the ranges used—so for the case shown, the power input would be limited to between 0.03 and 0.26 W. Although at first this procedure limits the authority that a user may have over input power for high-speed control applications, one can easily imagine a controller that relaxes the range limitations for short periods of time while tracking error is large. Also, not allowing the wire to overheat or completely cool increases response time.

This method is applied to closed-loop control of the same flexure system used in characterization. The coordinate frame is established such that $\delta = 0$ when $\varepsilon = \varepsilon_{pre}$.

5.5.2 Feedback Control Scheme

Once the mapping coefficients in Eq. (5.3) and the power ranges are obtained, the mapping can be employed in a simple feedback controller. The block diagram in Fig. 5.16 shows how the Joule heating power and electrical resistance that are measured by the power controller are passed into the mapping algorithm. Then the deflection, δ, is commanded, and a simple PI controller calculates the command power for the next loop. The PI gains were tuned manually and then maintained for all experiments, but no significant effort was made to optimize them. Additional control features, such as a feed-forward algorithm, a mechanism to relax input power range

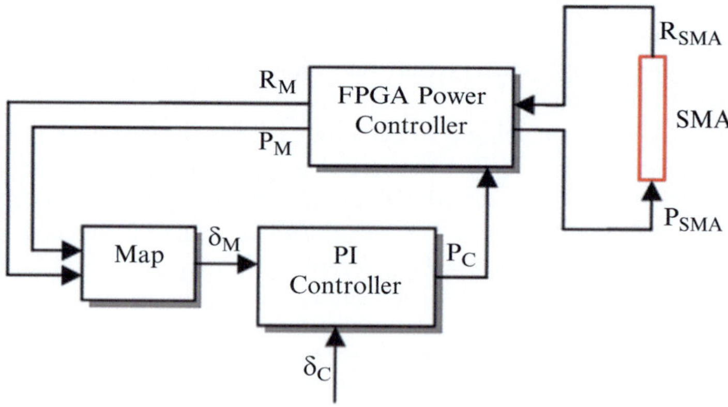

Fig. 5.16 Controller and mapping block diagram

limitations for a short period of time while tracking errors are large, or model-based controller could be employed. However, the primary focus is to analyze the baseline performance of the mapping under the assumptions mentioned.

5.6   Results of Single SMA-Flexure Control

The performance of the controller and the accuracy of the resistance-based position feedback measurement were tested in response to sinusoid inputs at 0.2, 1.0 (not plotted), and 2.0 Hz as well step inputs. Figure shows that at 0.2 Hz, the controller is able to force the SMA to follow the command positions very well. Tracking error, which is equal to the difference between the command position, δ_C, and the resistance-based position measurement, $\delta_{M,\text{res}}$,

$$E_T = \delta_C - \delta_{M,\text{res}} \tag{5.4}$$

is 8 μm RMS (root mean square), as measured over the last ¾ of the time. The more relevant error for this study comes from the measurement. The measurement error, calculated by subtracting the resistance-based displacement measurement from the laser-based measurements, $\delta_{M,\text{laser}}$,

$$E_M = \delta_{M,\text{laser}} - \delta_{M,\text{res}} \tag{5.5}$$

is 21 μm RMS. For measurement, peak to peak ("PP") error is important because it represents the accuracy of the sensor. For this system, peak to peak measurement error is less than 103 μm, or 17.2 % of full scale. These metrics are labeled on the "Error" plots in Figs. 5.17 and 5.18. These plots show that measurement error has a tendency to change sign each time the displacement reverses direction. This is not surprising, because the resistance vs. strain characteristic in Fig. 5.17 has a slight hysteresis. This hysteresis affects the measurement accuracy every time the phase transformation process switches direction.

The right plots in Figs. 5.17 and 5.18 show the SMA resistance plotted vs. displacement as measured by the laser (red) and the resistance-based mapping (blue). The mapping shows up as a straight line, since mapped displacement is calculated directly from the resistance measurement via the slope and intercept of the best-fit line. In this implementation, the linear approximation is very effective.

When the input frequency is increased to 2.0 Hz in Fig. 5.18, a noticeable lag can be detected, and the actual displacement does no longer follow the set point perfectly anymore. The measurement error is 22 μm RMS, 73 μm peak to peak, and it still changes direction with each cycle; also the lag now causes a very large tracking error. This is due to the fact that the controller has reached its limits and runs into saturation as evidenced by the power signal in the bottom row of the figure. However, the laser position signal and the position reconstructed by the resistance-based self-sensing algorithm are still in very good agreement, clearly demonstrating the feasibility of the self-sensing concept.

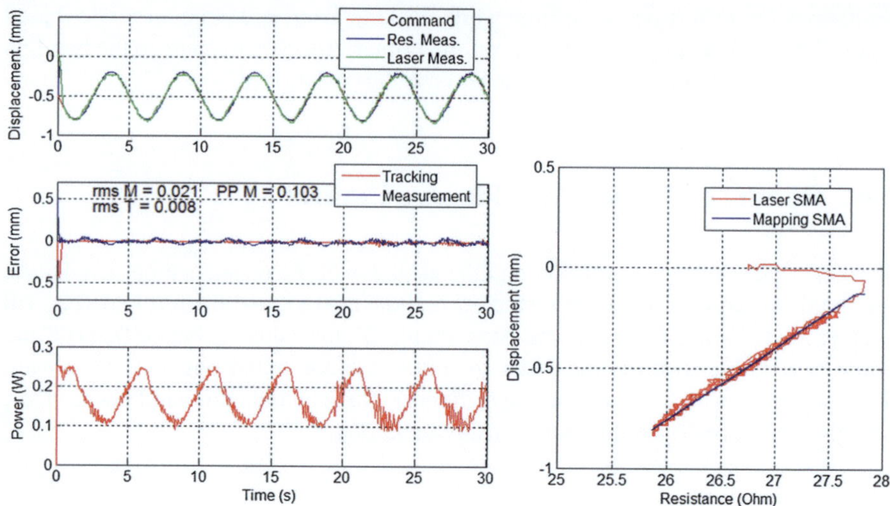

Fig. 5.17 Position tracking of 0.2 Hz sinusoid (*left*) and sensor diagram showing displacement as measured by the laser and the R–δ mapping (*right*)

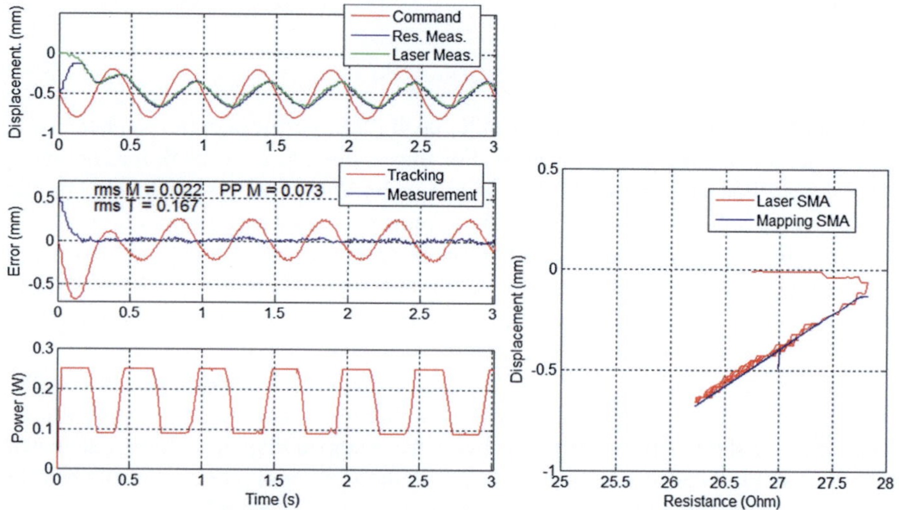

Fig. 5.18 Position tracking of 2.0 Hz sinusoid (*left*) and sensor diagram showing displacement as measured by the laser and the R–δ mapping (*right*)

References

1. V. Novak et al., Electric resistance variation of NiTi shape memory alloy wires in thermomechanical tests: experiments and simulation. Mater. Sci. Eng. A **481–482**, 127–133 (2008)
2. S.J. Furst, J.H. Crews, S. Seelecke, Numerical and experimental analysis of inhomogeneities in SMA wires induced by thermal boundary conditions. Continuum Mech. Therm. **24**(4-6), 485–504 (2012)

3. S. Seelecke, N. Lewis, *Effects of Temperature Boundary Conditions on SMA Actuator Performance Using a Fully Coupled Thermomechanical Model*, in *Proceedings of SMASIS conference*, Scottsdale, AZ, USA, 2011, p. 487–492. ISBN: 978-0-7918-5471-6
4. S.J. Furst, R.Hangekar, S. Seelecke, *Practical Implementation of Resistance Feedback Measurement for Position Control of a Flexible Smart Inhaler Nozzle*, in *Proceedings of the ASME Conference on Smart Materials, Adaptive Structures and Intelligent Systems*, 2010, vol 2, p. 205–213

Chapter 6
Potentials of Shape Memory Technology in Industrial Applications

Sven Langbein

The application potentials of shape memory components are much more diverse than those of other function materials. They range from actuator applications to sensors and passive damping systems.

This range of possible applications exists thanks to the various industrially usable effects. The mechanical effect with its special damping properties that occurs in shape memory alloys expands the field of possible applications in particular. Such passive elements are easy to integrate into existing systems. Unfortunately, this effect has not yet become prevalent in industrial applications so far. Only in medical technology has the effect been applied intensively. In industrial applications, active elements on the basis of the thermal effect are more commonly used. Figure 6.1 shows once again the mapping of applications to the effects of SMA. The sensory applications play an important role here, because they can operate on the basis of both effects. This depends on the type of variable to be detected.

Shape memory technology is a cross-sectional technology due to the diversity of its applications. The usability of SMA elements across industrial applications is very diverse. A variety of both active and passive components can be generated with the material. To illustrate the diversity of applications, several technical application potentials are listed again below. However, the applications mentioned here represent only a limited selection:

- Electrically activated actuators with linear travel motion as substitution for electromagnets, e.g. in unlatching systems.
- Thermally activated actuators as substitution for expansion elements, e.g. as drives in thermostatic valves.
- Systems that compensate temperature-induced changes in the viscosity of fluids, e.g. for hydraulic systems.

S. Langbein (✉)
FG-INNOVATION GmbH, Universitätsstr. 142, 44799 Bochum, Germany
e-mail: Langbein@fg-innovation.de

© Springer International Publishing Switzerland 2015
A. Czechowicz, S. Langbein (eds.), *Shape Memory Alloy Valves*,
DOI 10.1007/978-3-319-19081-5_6

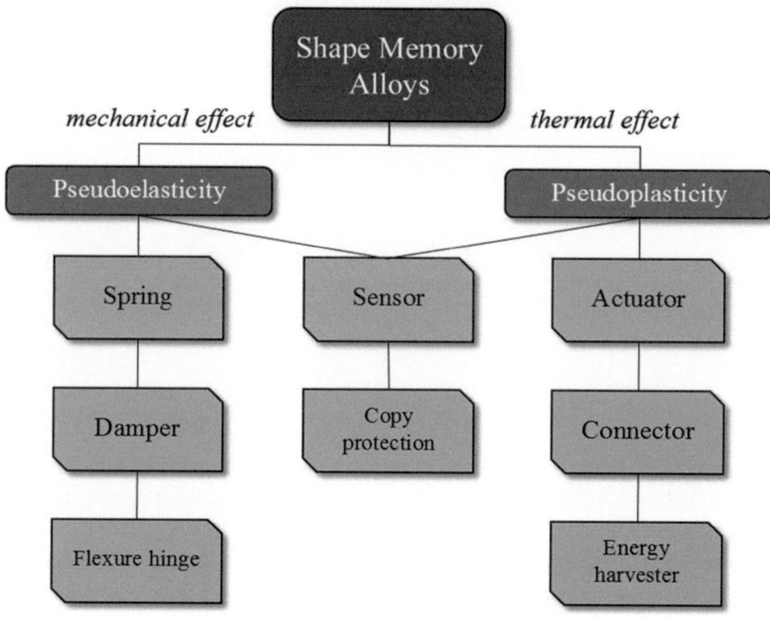

Fig. 6.1 Application potentials of SMA components subdivided by material effects

- Systems that compensate thermal expansions, e.g. to implement backlash-free gears.
- Systems that recover energy from flowing gases or fluids at temperature differences of only a few Kelvin.
- Thermally activated connecting elements as substitutes for screws, rivets, or plastic clips.
- Compact spring elements with almost horizontal spring characteristic, as a substitute for disc springs, e.g. for use in tool clamping systems.
- Highly elastic elements, e.g. for use as solid joints.
- Vibration damper with SM effect, e.g. for use in machinery or buildings.

6.1 Actuators

Shape memory effect-based actuators belong to the group of material-mechanical actuators. The term "material-mechanical" illustrates the transducer principle and the underlying physical effects or operating principles, respectively. Furthermore, these material-mechanical effects are inverted, so that actuators from this group can also be used as sensors. Electrically activated SMA actuators consist usually of an electronic controller, the actual material-mechanical transducer, and sometimes a downstream converter. Unlike conventional drive principles, the use of SM-based actuating elements offers the possibility to produce simply constructed, lightweight,

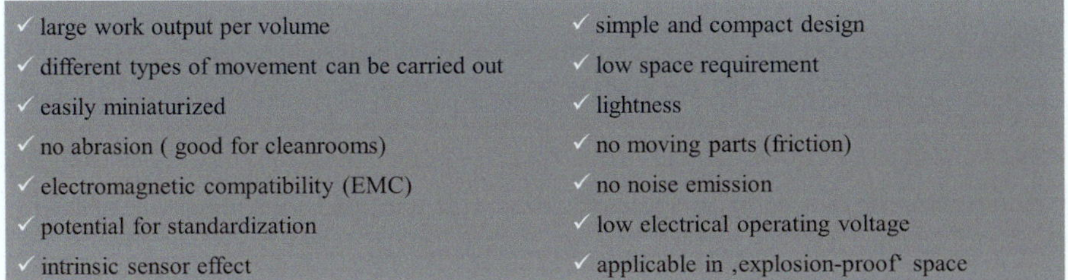

Fig. 6.2 Benefits of SMA actuators with respect to industrial applications

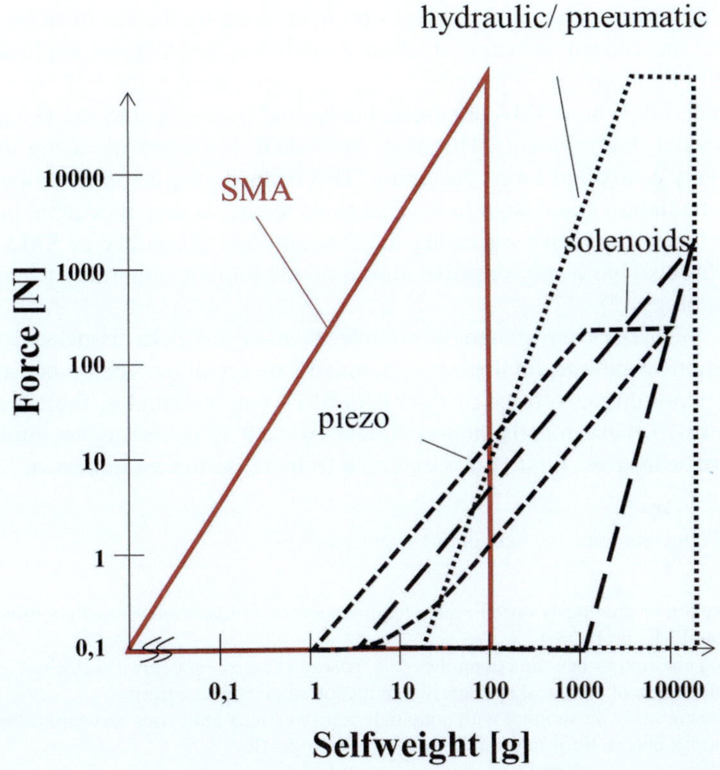

Fig. 6.3 Self-weight with respect to the positioning force of different actuator principles [1]

and noiseless actuators. The key advantages of SMA actuators are summarized in Fig. 6.2.

Figure 6.3 further illustrates the main advantage of SMA actuators—the low weight in relation to the positioning force—compared to other actuator principles. Visibly significant here is the extremely high weight-specific performance of SMA actuators.

Despite all these advantages, the potential of this technology can only unfold in its entirety by using standardized components. The following chapters outline why that is.

A disadvantage of SMA-based actuators, however, is insufficient dynamic response of certain applications. Other problems, such as low transition temperatures and stability characteristics, are further topics of current research. Product requirements should be reviewed especially with regards to these critical areas for all development projects.

6.1.1 *Opportunities and Risks*

Despite the many technical advantages brought along by the use of shape memory actuators, the current industry situation is such that SMA-based applications are practically non-existent in manufacturing. A main reason is the lacking awareness of the material. Where SMA are used in industrial products, they are usually based on individual developments. However, individual developments using this material are very costly and time consuming. This is due to the complex properties and lack of simulation and design tools. Extensive testing is thus inevitable in order to make definite statements regarding the function and durability of SMA components. This also has a very negative impact on the implementability of shape memory technology [2].

Also, businesses are generally apprehensive to the risks associated with the introduction of new technologies. A summary of the above-mentioned risks, but also the opportunities offered by the use of SMT, can be found in Table 6.1.

To resolve these mostly non-technical risks, it is necessary to implement a variety of strategies. Firstly, it is essential to increase the awareness of SMA and

Table 6.1 Opportunities and risks of SMA actuators

Chances
• Resolution of previously unachievable requirements (e.g. noiseless drives, light drives, drives without EMC problems)
• Implementation of new functionalities (e.g. position control or status monitoring)
• Simplification of technical systems and reduction of electronic efforts
• Interdisciplinary technology with potentials across various industries and applications
• Enormous potential for innovation, particularly for SMEs
• Significant improvement in material quality of SMA in recent years

Risks
• Development tools for SMA-based components are missing
• High development effort for individual developments, lack of standardized actuator systems
• Many trial-and-error experiments needed with varying parameters to achieve optimal results
• Sensitivity of functional properties to environmental factors, such as thermal or mechanical loads
• Consistent quality standards and norms for semi-finished products are not yet available
• Lack of knowledge and experience with SMA in industrial enterprises
• Lack of risk-taking with regard to the implementation of new technologies
• Small number of material suppliers and actuator producers

to clarify the technical potentials of the material. Furthermore, the provision of standardized drives on the market is an efficient way to reduce the risk and expense of companies, which is crucial in the implementation of the shape memory technology. The risk will then be borne by the drive manufacturer, but can be reduced due to their knowledge and experience. The form such standardized drives can take and which application potentials can be tapped into is outlined in further detail throughout the following chapters.

6.1.2 Application Potentials

Figure 6.4 schematically illustrates the potential field of SMA drives. A distinction is made here between the activation type of the SM effect, which also includes the usage potential of the drives.

Firstly, drives that are thermally activated by their environment should be mentioned. Such drives are used, for example, in thermostatic valves in heaters [3]. This type of drive is found rather commonly on the market due to its simple structure. In most cases, it is not technical problems that limit implementation, but cost. Such SMA drives compete with expansion elements or thermostatic bimetals, both of which can be purchased at low cost today. With regard to those drives, the advantage of SMA-based systems lies in their compact structure and better dynamics. By using the R-phase, SMA applications with a hysteresis of only 2 K can be generated. Such SMA-based drive have already been used with success in faucets and heating systems. A shower fitting based on the R-phase technology has already been

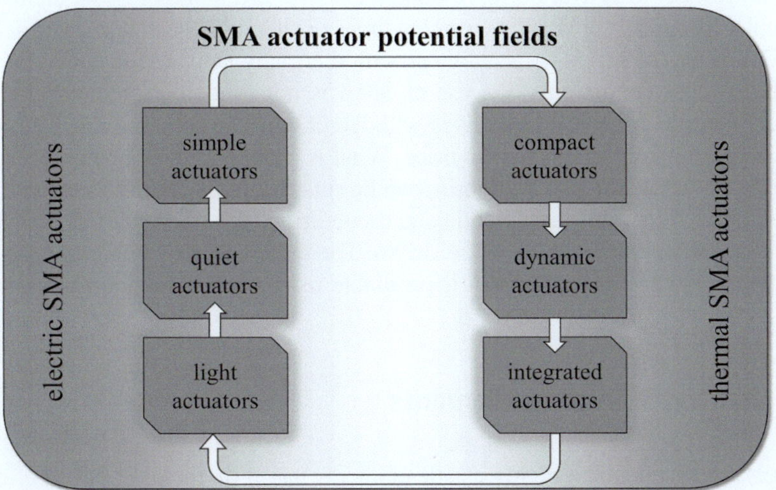

Fig. 6.4 General application potentials of SMA in the field of actuator technology

Table 6.2 Substitution potentials of SMA actuators

Substitution potentials of thermal wax elements and thermostatic bimetals by shape memory actuators	*Substitution potentials of solenoids and electric motors by shape memory actuators*
Electrification of applications (external regulation by the user or controllers)	Back to the roots: de-electrification or simplification of applications (autonomous control by the ambient conditions)
Redundant operation is in the foreground (combination of thermal and electrical activation)	Weight reduction is in the foreground (e.g. aircraft and cars)
Multifunctionality is in the foreground (integration of monitoring functions)	Silent operation is in the foreground (e.g. high-priced cars)
High forces or compact dimensions are in the foreground (e.g. miniaturization of valves)	Security aspects are in foreground (e.g. high-speed activation of security valves)
Specific component geometries and deformations in the foreground (e.g. special packaging specifications)	Space reduction is in the foreground
Prerequisites for substitution	
Cost requirements must be fulfilled	Dynamic requirement must be fulfilled
Load and stroke requirements must be fulfilled	Operating temperatures must be realizable

presented in [4]. In the future, new functionalities, such as sensory or monitoring functions, will also be integrated into such thermostat drives, which will lead to a boost in SMA application.

Then there are electrically activated systems. Such drives are used as control elements, e.g. for unlatching/unlocking functions in the interior of motor vehicles, and thus provide an alternative for electromagnets and electric motors. The advantages of SMA are the low weight, the noiselessness, and simple design. The dynamic behavior with respect to activation represents a decisive advantage of SMA elements. In particular for safety applications tripping times of a few milliseconds can be achieved [5]. Cost-wise the gap is not as large as with thermally activated systems. However, looking at cost factors only, the use of SMA is not profitable. It should be noted, though, that the price of SMA drives will decrease in the future, while electromagnets are more likely to increase in price. What is profitable today, however, is when not only the actor, but also the sensor, can be substituted by an SMA element.

Table 6.2 illustrates the potentials and conditions for displacement or substitution of conventional drives by SMA drives. The table offers simplified assistance in deciding whether the use of SMA is profitable from a technical point of view.

6.2 Spring/Damping Elements

Suspension systems and highly elastic elements based on the mechanical effect are nowadays mainly used in medical applications. Medical technology retains the largest market share also in relation to SMT. The main products today are stents.

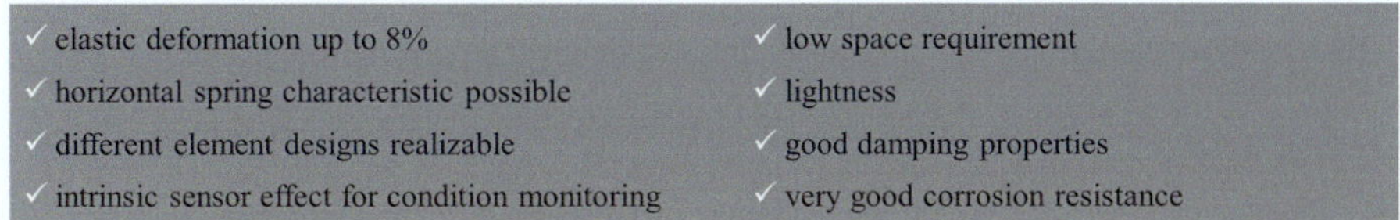

Fig. 6.5 Benefits of SMA springs with respect to industrial applications

However, the medical orientation of SMT has the disadvantage that the price of semi-finished products is maintained at a high level, and the transition to industrial applications is deterred. Since sufficient market penetration has already been achieved in the field of medical technology, there is less need for introduction of SMT, and this product range will not be pursued further in this book.

For industrial applications, however, the mechanical effect has enormous potential as well. In particular, the substitution of conventional springs, such as disc spring packages, can produce a variety of innovations. To highlight a few: the significant reduction of installation space and moving masses, as well as the possibility of realizing an almost horizontal spring characteristic. These and other crucial advantages of SM springs are summarized in Fig. 6.5. Despite all these advantages, the mechanical effect is even less commonly known in mechanical engineering as the thermal effect, so that a lot of educational work will be required.

A disadvantage for certain applications using shape memory alloys, for example with springs and dampers, is the not yet sufficient stability of the effect. The service life of several million cycles is significantly higher than that of thermal applications. SMA elements are therefore not suitable for damping of high-frequency vibrations. Another problem is the strong dependence of spring and damping characteristics on ambient temperature. This property of the mechanical SM effect has been covered in the introductory chapter. For certain applications, this property may be beneficial, for example wherever temperature-adaptive spring behavior is required.

6.2.1 Opportunities and Risks

Despite the many technical advantages that come with SMA springs, there exist barely any applications in the industrial sector comparable to SMA actuators. The main reason, again, is the lack of awareness of this extraordinary material. Aside from the advantages already mentioned, the use of SMA springs also carries the risk of having to perform extensive evaluation tests. In comparison to SMA actuators, however, more matured and better simulation tools exists, such as in the form of FEM systems. This is due to the fact that pseudoelastic behavior can be more easily simulated than the pseudoplastic behavior. In addition, the development of simulation methods for pseudoelastic components has been driven by medical technology for decades. A concluding summary of the main risks, but also opportunities, of the use of shape memory alloys with mechanical effect is presented in Table 6.3.

Table 6.3 Opportunities and risks of using SMA springs

Opportunities
• Resolution of previously unachievable industry requirements, e.g. horizontal spring characteristic or loadable solid joints
• Implementation of new functionalities into spring systems, e.g. condition monitoring or damping behavior
• Variety in application potential, particularly in clamping and connection technology
• Enormous potential for innovation, particularly for SMEs
Risks
• Experiments with varying parameters for product evaluation necessary
• Sensitivity of the functional properties to environmental factors, such as ambient temperature
• Lack of knowledge and experience with SMT, particularly with regard to the mechanical effect in industrial enterprises
• Lack of risk-taking with regard to implementation of new technologies
• Small number of material suppliers and monopoly of medical technology
• High cost of semi-finished products, such as wires and sheets

As with actuator applications, it is essential to increase the publicity of this material effect, and to clarify the technical potentials of pseudoelasticity. Also in this case, offering standardized spring elements on the market would be a promising opportunity to move companies to use this technology in their products. Through standardization, the experimental effort, for example, would be significantly reduced. The price would also drop due to larger purchase quantities.

6.2.2 Application Potentials

So far, the mechanical SM effect has not yet prevailed in industrial applications. Despite the excellent damping behavior, which enables the implementation of various passive damping systems, applications based on this effect are an absolute rarity. There is, however, a current trend toward compact spring systems. For example, considerations are taking place as to whether disc springs in tool clamping systems should be replaced by SMA-spring components. In this case one would utilize (a) the ability of pseudoelastic SMT to generate an almost horizontal spring characteristic, and (b) the ability to generate very compact spring elements with highly reversible deformability. One would reduce the overall installation size, as well as the moving masses, in the field of clamping systems. Ultimately, this leads to a reduction of required labor and would also improve processing quality. The mechanical effect has an enormous application potential not only in clamping technology, but also in other fields, for example in connection technology. Pseudoelastic washers or bolts can prevent relaxation processes. With such applications, conventional systems are not only substituted, but completely new areas of application and technical functionalities are created. The implementation of sensory abilities in SMA-based spring systems continues to be an important goal. Using the internal sensor effect, one can also monitor SMA springs in a variety of ways. Functions such as position detection or a service life and fatigue monitoring should be named here. But changes and adjustments of spring and damping characteristics during operation are also possible with SMT. Thus, spring elements that were previously used in a purely passive manner are now entering a new level of technology.

6.3   Sensors

SMA-based actuator systems offer various advantages over conventional drives. In particular, the integrated sensor function has a special role in aiding the SMA-drives success. This enables us to generate actuators that are technically simple, yet complex in functionality. Conventional displacement sensors can thus be dropped, which has very positive effects on the cost balance of an actuator as well.

This inner sensor effect is caused by the significant change in electrical resistance during the phase transition. The electrical resistance of a conductor can be easily read nowadays, which is why sensor function does not constitute a particularly complex problem in electronics. Furthermore, it adds positively to the sensory capabilities that the phase transition is very sensitive to external and internal influences, which is reflected in the resistance curve and can be read thus. Such influences are, for example, external load conditions and ambient temperatures, but also internal factors like the effects of fatigue. This section is intended only as an introduction, since the sensor functions of SMA elements in connection with controlling of SMA actuators are explained already in Chap 5.

The sensor function does not necessarily require the combined use with an actuator element, but it offers a large application potential as a separate function as well. This new category of pure SMA sensors is currently still a research topic, but will become increasingly important as demand for new sensor systems is significantly greater than the demand for new drive systems. Since pure SMA sensors usually operate on the basis of mechanical SM effect, a combination of SMA-based sensors with SMA-based spring and damping systems, similar to that of actuators, is also possible. A more detailed look at the advantages and disadvantages is not possible at this stage because of the diversity of sensor principles. Only after cases of specific uses compared with the prevailing principles of sensors there, we will be able to provide a more detailed analysis.

6.3.1   *Opportunities and Risks*

In contrast to actuator applications of SMA elements, the use of SMA as a pure sensor is still largely unexplored. Before industrialization can take place, several issues are still to be addressed through basic experiments. For this reason, the use of SMA sensors is also associated with a greater risk than the use as an actuator or spring element.

Table 6.4 Opportunities and risks of using SMA sensors

Opportunities
• Development of innovative and robust sensor systems
• Implementation of new functionalities in technical systems, e.g. condition monitoring or overload detection
• Wide range of usage potential in the form of disposable sensors
• Very good chances to obtain intellectual property rights, as this area is not currently a focus of the industrial sector

Risks
• Basic research is required
• Sensitivity of the electrical resistance curve to environmental factors
• Identification of application fields required
• Lack of risk-taking with regard to the implementation of new technologies
• Small number of material suppliers and monopoly of medical technology
• High cost of semi-finished products, such as wires and sheets

Since pure SMA sensors work on the basis of the mechanical effect, the already existing simulation tools can contribute to the prediction of the pseudoelastic behavior, for example, in the form of FEM systems. The reading of sensory information from interpretations of the resistance curve, however, still represents new territory. A concluding summary of the main risks, but also the opportunities, of the use of SMA with mechanical effect as sensors is presented in Table 6.4.

6.3.2 Application Potentials

Industrial applications of pure SMA sensors do not exist in practice thus far. The sensory attribute is currently used only in conjunction with actuator tasks. However, the theoretical potential for application of SMA sensors is large. In addition to use as power or position sensors, the use as a fatigue and wear detection sensor, or for the detection of excessive load conditions, would make sense as well. This means tapping into completely new fields, for which no sensor previously existed. An exemplary application, which is already legally protected, is the collection of sterilization cycles for medical devices. Other applications in this context will surely follow in the near future. The realization of disposable sensors, for single-use only, is possible with SMT, because the sensor can be as simple as a piece of wire, which can possibly even be read externally.

References

1. S. Langbein, A. Czechowicz, *Konstruktionspraxis Formgedächtnistechnik* (Springer Vieweg Verlag, Mannheim, 2013). ISBN 3834819573
2. S. Langbein, K. Lygin, T. Sadek, Significance of requirements for the implementation of new technologies using shape memory technology. in *Proceedings of the 18th International Conference on Engineering Design (ICED)*, Copenhagen, Denmark, 2011

3. M. Humburg, G. Eggeler, M. Wagner, *Thermo-Kombiventil: Thermomanagement im Standheizbetrieb. Automobiltechnische Zeitschrift (ATZ) 03/06, Jahrgang 108* (Springer Vieweg Verlag, Wiesbaden, 2006)
4. I. Ohkata, Y. Suzuki, The design of SMA actuators and their applications, in *Shape Memory Materials*, ed. by K. Otsuka, C.M. Wayman (Cambridge University Press, Cambridge, 1998)
5. A. Czechowicz, J. Boettcher, S. Mojrzisch, S. Langbein, High speed shape memory alloy activation. in *Proceedings of SMASIS Conference*, Stone Mountain (Georgia), USA, 2012

Chapter 7
Shape Memory Valves: Motivation, Risks, and Potentials

Alexander Czechowicz and Sven Langbein

7.1 Introduction and Classification of SMA Valves

Shape memory alloys (SMA) exhibit the remarkable property to recover a previously imprinted shape after a deformation. Shape memory alloys have certain characteristics which are unique in comparison to other actuating principles, mainly the high mechanical stress generation potential. Furthermore shape memory actuators make noiseless actuation possible, which is desired in automotive comfort applications. Since years shape memory alloys have been used as thermal and electrical drives for valves. These smart materials show remarkable benefits in comparison to solenoid and electromotor drives like noiseless actuation, lightweight design, compact design, and low electromagnetic noise potential. Generally, SMA valves can be distinguished by their purpose as control applications, safety applications or both as shown in Fig. 7.1. An other possibility is to classify valves by their activation principle. This classification is shown in Fig. 7.2. Here we can distinguish a thermal activation, an electrical activation or a combination of both.

While cyclic SMA valves have often to switch up to 500,000 [1] times their state from martensite to austenite, the actuation in safety application is needed only one or little times in emergency cases. Both SMA valve drives can be triggered by thermal fields or electrical currents. With closer look on valves for cyclic applications, application scenarios for thermal and electrical SMA valves can be distinguished too. Serial devices, like shown in Fig. 7.3, can be found today in automotive and industrial

A. Czechowicz
Zentrum für Angewandte Formgedächtnistechnik, Forschungsgemeinschaft Werkzeuge und Werkstoffe e.V., Papenberger Straße 49, 42859 Remscheid, Germany
e-mail: czechowicz@fgw.de

S. Langbein (✉)
FG-INNOVATION GmbH, Universitätsstr. 142, 44799 Bochum, Germany
e-mail: langbein@gmx.com

© Springer International Publishing Switzerland 2015
A. Czechowicz, S. Langbein (eds.), *Shape Memory Alloy Valves*,
DOI 10.1007/978-3-319-19081-5_7

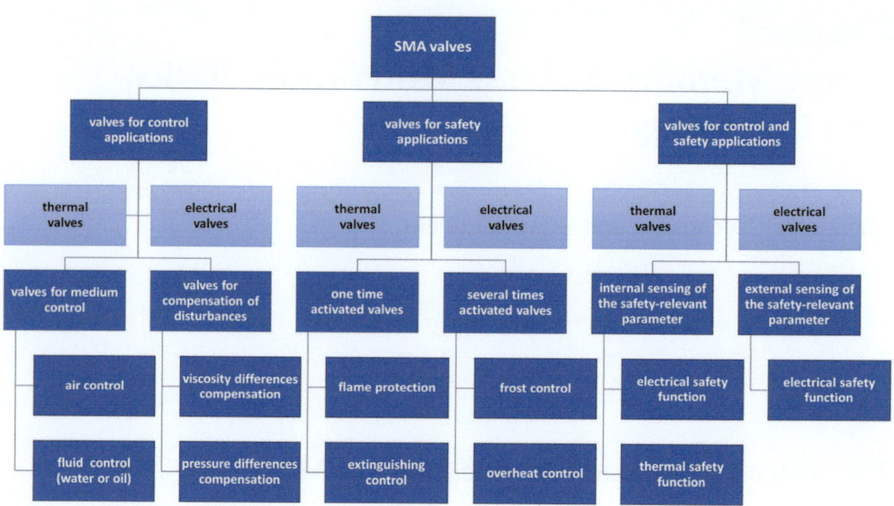

Fig. 7.1 Characterization of SMA valve application scenarios

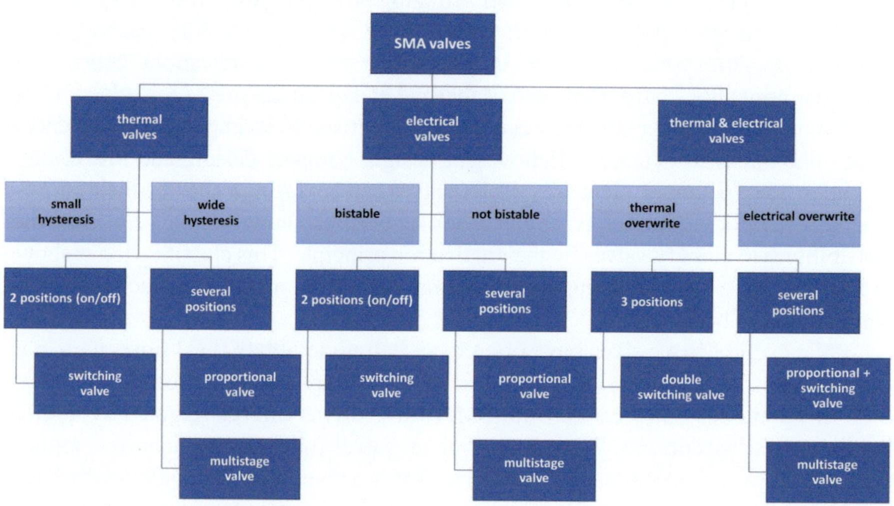

Fig. 7.2 Examples of SMA valves on market [2, 3, 5]

applications. Figure 7.3a shows an electric SMA valve system for the control of air pressure in automotive seat systems [2], while Fig. 7.3b presents an electric microvalve for medical equipment [3]. In Fig. 7.3c two valves using low hysteresis SMA [4] for thermal regulations in home heating systems like underfloor heating systems are presented [5]. These applications are explained in-depth.

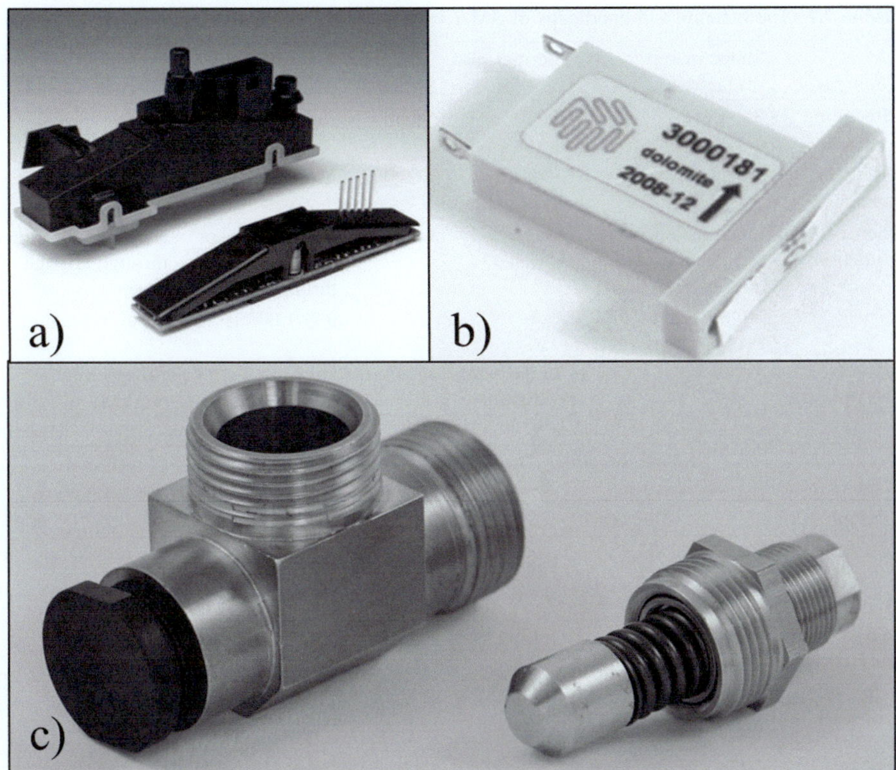

Fig. 7.3 Comparison of alternative smart valve principles

7.2 Benefits and Handicaps of SMA Valves

Regarding the potentials of the described SMA valves, several benefits and handicaps of the systems can be summed up as in Table 7.1. The comparison has been made for competing thermal and electrical systems for cyclic valve application. Due to their high energy density, SMA actuators can provide high actuating forces which can be either used for directly actuated normally open or normally closed valves. Such valves are determined by simple design, without the need of a pilot valve system. This is a major benefit, because lifetime of conventional valves is often determined by the lifetime of components of pilot valves (e.g. membranes and fittings) and not by the lifetime of the actuating solenoids [6]. Another major benefit is the intrinsic sensor function, which allows detecting the SMA's stroke, overload states, and the fatigue [7, 8].

In comparison with thermostatic wax actuators [5], thermal SMA valve drives are able to react faster on thermal field variations due to their remarkable smaller material volume at equal force output level. The mass ratio of SMA to wax actuators lies between 10 and 30 % with a larger surface–volume ratio [3]. As a result, the heat exchange between the fluid and the actuator is much faster.

Table 7.1 The benefits and handicaps of SMA valves in comparison to competing valve drives

Benefits	Shape memory actuator	Thermal wax element	Thermal bi-metal	Electric piezo-actuator	Electric solenoid
Installing space	Smallest	Medium	Small	Small	Large
Actuator weight	Very low	Medium	Low	Medium	High
Sensor functions	Intrinsic sensor elements, usable for electrical and proportional valves	None	None	Intrinsic sensor elements through piezo-resistivity	Mechanical state through voltage peaks
Actuators' complexity	Simple	Complex in production	Medium	High (electronics)	Complex in production

Handicaps	Shape memory actuator	Thermal wax element	Thermal bimetal	Electric piezo-actuator	Electric solenoid
Fatigue [cycles]	<1,000,000	>1,000,000	>10,000,000	>5,000,000	>5,000,000
Cyclic dynamics in el. applications [Hz]	<10 Hz	<0.1 Hz	–	<10 kHz	<1 kHz
Costs (compared at 10,000 pcs/year)	Medium	Low	Low	High	Medium
Max. ambient temperature [°C]	140 °C (safety) 80 °C (cyclic)	110 °C	350 °C	180 °C	200 °C
Customer's familiarity with the actuator principle	Low	Medium	Medium	Medium	High

While pulling stress levels can achieve more than 450 N/mm^2 [9], the dynamic response of SMA is much slower than solenoid and piezo drives. This does mainly be of concern for cyclic actuations, because SMA can be triggered within less than 0.5 ms [10]. The deactivation is determined by thermal convection which is responsible for the cooling and retransformation speed of SMA elements. Even if this reduces the application spectrum of SMA valves for high-frequent control valves, still SMA can be used in safety applications, focusing technical additional values like "non-corrosive drive material." Other handicaps are the lower life-cycles [9] and lower operating temperatures [3, 11]. Despite these disadvantages, there are still several promising SMA valve applications on the market. A survey made by the authors of 51 valve manufactures in Germany and France resulted in following economical conclusion: One of the major handicaps of SMA technology is the lower economical efficiency compared to the other actuating principles. In comparison with well asserted actuator systems like thermostatic wax elements for thermal, and solenoids for electric valves, the cost of SMA systems is often higher. From a strategic point of view, there are hardly motivations for valve producers for the utilization of SMA valves if the customer is not interested in size, noise and weight

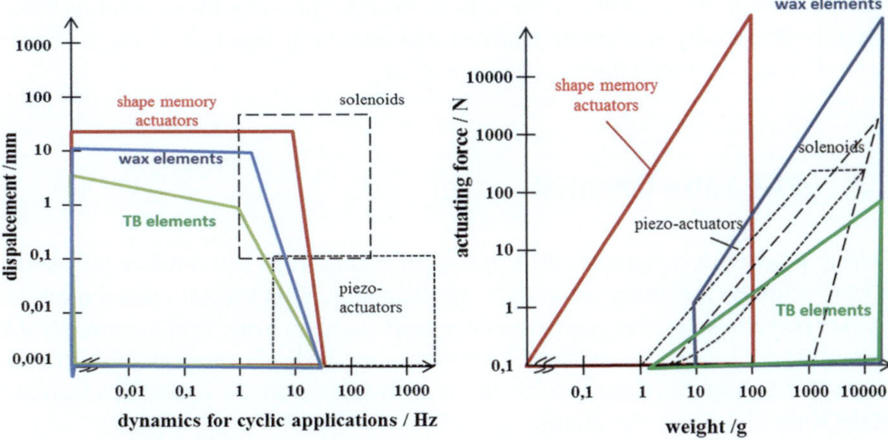

Fig. 7.4 Comparison of a solenoid and an SMA valve principle

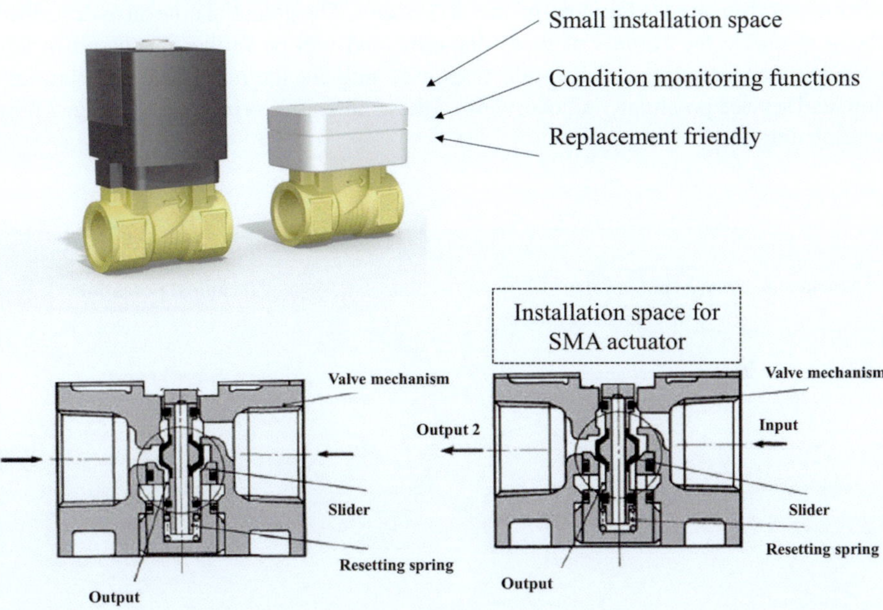

Fig. 7.5 The scored potential criteria of SMA valves sorted in the potential classes: technical, production, and market potentials. The *black line* shows exemplary how the rating can be noted in the diagram

reduction as often in machining or process industry where valves are sold on mass markets. A comparison of valve drives is shown in Fig. 7.4.

The advantages of SMA exemplary for a water valve are shown in Fig. 7.5. An electrically activated SMA element competes in this case with an electromagnet. The solenoid is significantly larger and heavier than the SMA actuator. The main advantage for industrial applications, however, provides the monitoring functionality. Especially for remote maintenance and the integration of the valves in central process controls this function is crucial.

7.3 SMA Valve Potentials

During product development, the methodical approach as presented in [12] starts always with the definition of product specifications. With special attention on the market potentials, it is necessary to distinguish the main benefits of a certain SMA valve concept for specific markets. Therefore, a methodical overview over different concepts has been developed within the competence centre for hydraulic machines at the Ruhr-University Bochum.

SMA-driven valves can be categorized with aid of evaluation diagrams which concentrate on the potential classes: technical potentials focusing on SMA-specific characteristics, production potentials focusing on product cost and time, and the in-use advantages compared to the state-of-the-art systems. The potential criteria as described above resemble the corners of these diagrams and will be further discussed in this overview concentrating on dynamic response, ambient thermal range, miniaturization, and service potentials. An overview of these potentials is presented in Fig. 7.6 by a rating diagram.

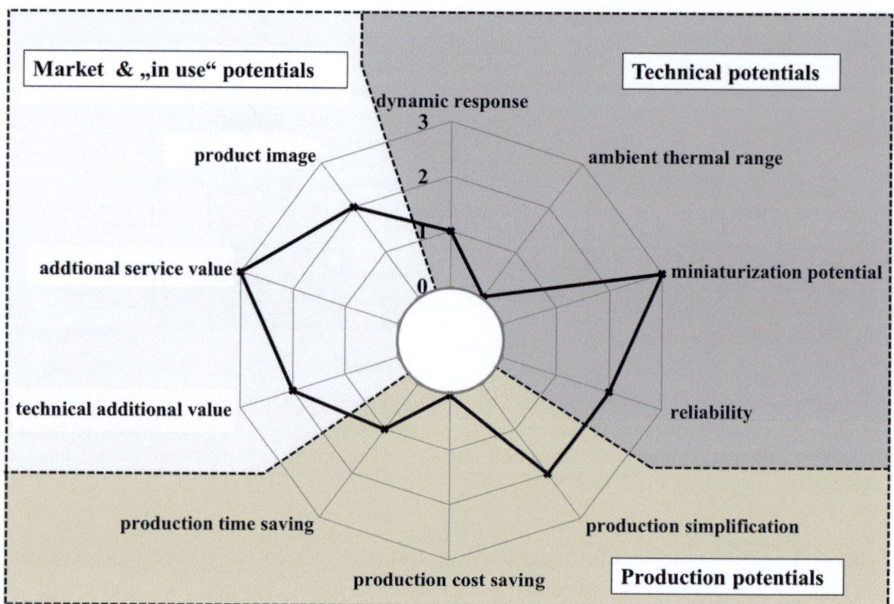

Fig. 7.6 Schematic overview of service applications with SMA components in valves

These criteria compare SMA valve scenarios to the state-of-the-art systems on the market (solenoid and wax elements mainly). Generally the criteria have high, middle, or low potential. The center circle of the evaluation diagrams can be interpreted as "no benefits in comparison to the state of art," while future work will also integrate handicaps in comparison to on-market solutions. The definition of the potentials depends on interpretation issues and is determined in this publication in excerpt as follows:

7.3.1 Dynamic Response

The dynamic behavior of SMA depends mostly on its dimension and the ambient conditions [13]. While thin SMA wires can operate in frequencies up to 5 Hz, SMA coil spring actuators need far more energy input for actuation and a longer time for resetting. In comparison to solenoid drives, only thin SMA wires are able to compete regarding dynamic issues. In general, the dynamic response consists of the main factors like type of activation (electrical, indirect electrical, or thermal), energy level input, the utilization type (safety or cyclic system), and the cooling time (dependent on the heat transfer). The result is either the triggering time or the cyclic actuation frequency.

7.3.2 Ambient Temperature Range

This criterion describes the actuating potential for SMA applications at a certain ambient temperature (Ta). If the ambient temperature is lower than the Mf temperature, a complete retransformation, needed for cyclic applications, is ensured. Other scenarios referring to Ta involve SMA safety valves with no need for retransformation. These can work up to the austenite start (As) temperature. In example: many solenoid and wax system are capable of cyclic actuation at $T_a = 100\,°C$. Cyclic SMA applications are not possible at this temperature, while safety solenoid systems sometimes do not work at 120 °C. A high pre-stressed SMA actuator could work as a one-time safety application at this temperature. In analogy to the dynamic response, the potential criterion "ambient thermal range" consists of the main factors utilization type (safety or cyclic system) and the expected fatigue. In future, new alloy types can solve the temperature problem.

7.3.3 Miniaturization

This criterion describes the actuator volume and mass in applications. This affects not only the design space reduction, but as well a mass decimation of the valve drive itself. The miniaturization potential itself depends on the SMA output power. While

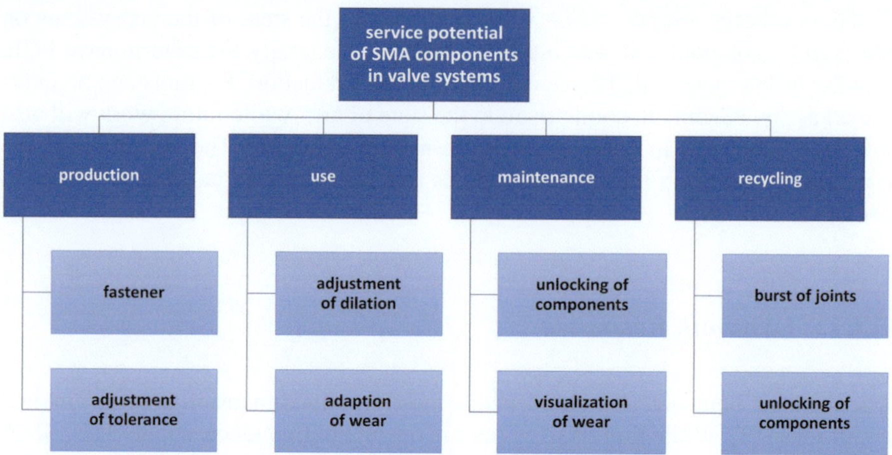

Fig. 7.7 Example of different SMA valve concepts with simplified potential scoring

thin SMA wires can generate relative high forces, the displacements are limited to maximal 8 % of the wires' length. SMA coil springs transform the mechanical energy commonly in increased displacement and lower actuating forces. In summary, the miniaturization depends mainly on the SMA element type and its dimension. In general, the miniaturization potential is high.

7.3.4 Reliability

This criterion describes the fatigue behavior, the durability, and the chemical resistance. The fatigue is a very complex attribute. It depends on a lot of parameters like the stress and the strain but also on the method of the integration of the SMA component in the system. In contrast, the chemical resistance is not a problem. Since SMAs are enormously resistant to typical corrosion types in industrial fields, they can be exposed to various media (commonly: water, oil, and compressed air). The activation of SMA drives in media (thermal activation) is a benefit given on electromagnets.

7.3.5 Additional Service Value

This criterion describes the potential of SMAs in service applications. Here, we can distinguish services for SMA products or services SMA components are used for. Figure 7.7 shows a schematic overview of the service applications, which may be realized through the use of SMA components in valve systems. In contrast Fig. 7.8

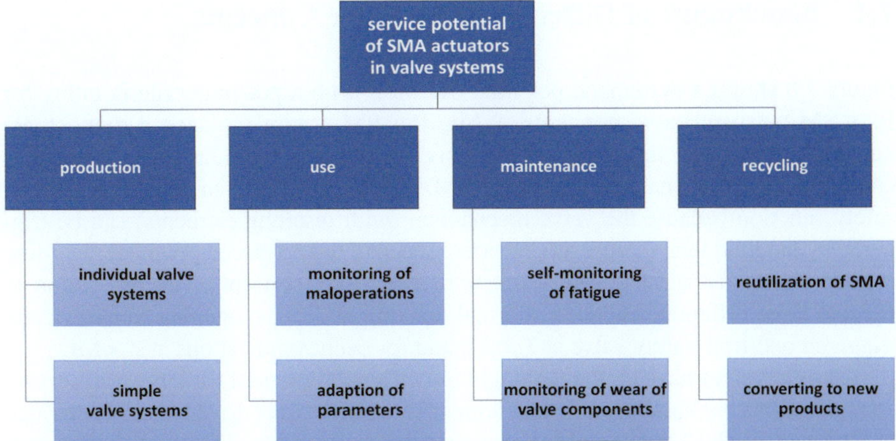

Fig. 7.8 Possible industrial product service systems for SMA valves in reference to the market cooperation between SMA-material and SMA-system provider

shows services that can be provided for SMA actuator systems. The change of paradigm from the separated view of product and services to a new product understanding consisting of industrial, integrated product-service systems (IPS[2]) is becoming increasingly important. The integrated view establishes innovation potential to increase the competitiveness of mechanical engineering thus allows business models in which the customer use, Fig. 7.8 Schematic overview of services applications for SMA actuators in valves. e.g. in the form of highly available machines and not the selling of the machine, is of central interest. Industrial product-service systems can achieve over 50 % of the value creation on business to business product markets. Therefore, it is necessary to define the potential of SMA elements regarding such market considerations. The recent research on SMA in combination with product-service systems, as introduced in [14], focused on the utilization of the intrinsic sensor function as condition monitoring feedback for fatigue and maloperation. Another interesting service potential is the production of SMA-driven systems for individual solutions in smallest serial amounts by usage of fused deposition production technologies. These themes are deepened in the following chapters.

7.3.6 Additional Technical Value

This criterion describes special technical potentials of SMAs. One example is the intrinsic sensor effect: In this case, the sensor effect for detection of the displacement or of excessive load conditions is used. This particular feature leads to a further simplification of mechatronic systems by counter the loss of additional sensors. The intrinsic sensor effect also leads to a further weight reduction. Thus, this is a clear advantage in comparison to solenoid drives.

7.4 Benchmark of Different SMA Valve Concepts

Figure 7.8 shows a systematic potential overview with a potential criteria rating for four SMA-driven valve concepts. While the SMA wire replaces a monostable solenoid drive for cyclic dosing applications in concept 1, a similar valve (concept 2) can use the thermal characteristic of the SMA element to fulfill an additional thermal safety function because the valve mechanism (with resetting elements) can be triggered by the fluid temperature too. Hence, concept 2 has advanced potential regarding service due to its intrinsic safety function. Therefore, concept 2 has the option to execute three different functions: thermal opening at $T_a > As$, opening as electrically triggered one-time safety valve at $T_a < As$, and for cyclic applications at $T_a < Mf$.

Thermal valves have on the first sight only a small potential for industrial service considerations. Generally, an SMA element is used as direct drive element activated by ambient or fluid temperature to operate the mechanism. The modification and customer-specific production of the semi-finished SMA component resembles only elementary product service. The addition of electrical condition monitoring to concept 4 can be seen as service upgrade. With the proper design, and an advanced resetting system, it is possible to use the electrical resistance characteristic to identify fatigue as well as for electrical heat treatment and modification of the transformation temperatures.

7.5 Service Concepts for SMA Valves

Within the potential overview of SMA valves, the focus can be set on industrial product service concepts. As proven by surveys, the relative higher prices of semi-finished SMA components avert the assertion of SMA valves. Hence the evaluation of possible additional values for market applications has to be found, so customers will accept higher system cost by lower maintenance or modification cost in running systems. On the other side, new market potentials for SMA material suppliers can also be found by regarding service potentials. This additional value is common in general machine markets like machining or automotive industry. As an example, automotive producers earn with a mid-class car the main income with services from maintenance, finance, to recycle services. In 2008 the average income on machine market was about 15 % regarding product service systems, while only 3 % were achieved by direct product sale [15]. This motivates to apply these considerations on SMA valve markets as presented in Fig. 7.8. While today's business concepts only focus on the SMA material provider who sells semi-finished SMA components to the customer, other economical potentials are disregarded. Future's perspectives consider industrial product service systems on four levels: during the engineering and production, during the in-use phase (maintenance), after the usage phase (recycling), and quality issues. During the SMA valve production future works will focus on concepts for product individualization which can be achieved either by product upgrades for new functions or customer–individual designs with rapid-manufacturing technologies. This is presented in chapter 12.2. During the use phase, installed SMA valves can be checked by the electrical resistance characteristics for their condition. A replacement friendly design allows substituting a SMA valve drive easily without the need of demounting the whole valve system, as well as the attaching of additional sub-components for product modifications as presented in the experimental results. Reshaping and possible modifications during recycling have been already discussed in [16], while quality management as major service potential will be discussed separately in further works. Again with closer look on Fig. 7.9, the additional market potentials can be discussed. Instead of selling semi-finished products to the customer, SMA suppliers may cooperate with service-oriented SMA system providers to enable service friendly SMA valve drives. These drives can be modified thermally or electrically as presented in the experimental section, or checked for functionality by the usage of detection of the electrical resistance and the SMA drive's fatigue over internet and data-bus connections. If SMA drives will fail (either by fatigue or by malfunction) it can be detected replaced by the supplier consortium automatically. They can decide if the systems will be recycled or renewed.

References

1. A. Böhm (2013), SMA in Serienanwendungen, 2nd Newsletter of SMA-Netzwerk, p. 3, 2/2011, http://www.SMA-netzwerk.de. Accessed 1 Oct 2013
2. M. Köpfer (2013), SMA-Akturatorik, Newsletter of Alfmeier Präzision AG, http://www.alfmeier.de. Accessed 1 Oct 2013
3. Datasheet of micro shape memory alloy valve, part no. 3000181, by dolomite-microfluidics Ltd., http://www.dolomite-microfluidics.com. Accessed 1 Oct 2013
4. K. Lygin, H. Meier, A. Czechowicz, Using a R-phase SMA methodology to design an energy harvesting unit with tight temperature hysteresis, (3112). In *Proceedings of SMASIS2013 Conference*, 16–18.09.2013, Snowbird (Utah), USA, by ASME, 2013
5. S. Langbein, A. Czechowicz, *Konstruktionspraxis Formgedächtnistechnik* (Springer Vieweg Verlag, Mannheim, 2013). ISBN 3834819573
6. D. Will, N. Gebhardt, R. Nollau, D. Herschel, *Hydraulik* (Springer, Berlin, 2006), pp. 213–322. ISBN 10: 3540343229
7. H. Meier, A. Czechowicz, S. Langbein, Geregelte Formgedächtnis-Antriebssysteme mit Widerstandsrückkopplung. In *Proceedings of Mechatronikkongress*, VDI-Tagungsband, 2011, pp. 345–350, ISBN: 978-3-00-033892-2
8. A. Czechowicz, *Adaptive und adaptronische Optimierungen von Formgedächtnisaktorsystemen für Anwendungen im Automobil*, (Ph.D. Publication, Ruhr-University Bochum, 2012), Shaker, ISBN: 978-3-8440-1433-4
9. A. Coda et al., *SmartFlex NiTi Wires for Shape Memory Actuators, Journal of Materials Engineering and Performance, 18 (2009), 691–695* (Springer Verlag, New York, NY, 2009)
10. A. Czechowicz, J. Boettcher, S. Mojrzisch, S. Langbein, High speed shape memory alloy activation, (8213). In *Proceedings of SMASIS2012 Conference*, 19–21.09.2012, Stone Mountain (Georgia), USA, by ASME, 2012
11. H. Meier et al., *Smart Control Systems for Smart Materials, Journal of Materials Engineering and Performance, 20 (2010), 559–563* (Springer Verlag, New York, NY, 2011)
12. S. Langbein, A. Czechowicz, A Multi-Purpose Method for SMA Actuator Development, (3053). In *Proceedings of SMASIS2013 Conference*, 16–18.09.2013, Snowbird (Utah), USA, by ASME, 2013

13. H. Meier, A. Czechowicz, S. Langbein, Service Systems for Shape Memory Technology. In *ASME of SMASIS 2011*, 18–21.09.2011, Scottsdale (USA), pp. 419–425, ISBN: 978-0-7918-5472
14. T. Lorenz, *Maschinen- und Anlagenbau Studie: Servicegeschäft—passive Vermarktungspolitik trotz steigender Bedeutung* (Research Publication of MP Marketing Partner AG, 2008)
15. S. Langbein, A. Czechowicz, *Strategies for Self-Repairing Shape Memory Alloy Actuators, Journal of Materials Engineering and Performance, 20 (2001), 564–569* (Springer Verlag, New York, NY, 2011)

Chapter 8
Design of Thermal SMA Valves

Sven Langbein and Konstantin Lygin

8.1 SMA Springs: Thermal Actuator Elements

In principle, all types of shapes (wires, sheets, springs, or free-form elements), are suitable for actuator functions. In industrial applications, however, wire and spring actuators have prevailed due to several advantages. An SMA actuator in wire design can apply a very high mechanical normal stress of up to 600 N/mm^2, with a simultaneous normal strain up to 8 %. An SMA-spring actuator applies a lower mechanical shear stress of up to 300 N/mm^2, with a shear strain of max. 2 %. Compared to a wire actuator, a spring actuator has the advantage of engaging a much higher displacement in the same installation space, because its spiral shape contains an integrated transmission ratio. Disadvantages relating to actuating force can be compensated by a larger wire cross-section.

Regardless of the actuator element's design, a SMA actuator consists of at least one actuator element and at least one return element. The actuator element realizes the control function upon heating. The return element actions the reformation of the actuator element into its original shape during cooling. Figure 8.1 sketches principles of simple SMA-spring actuators after the two-way effect. The left picture shows how a constant restoring mass influence the behaviour of the SMA actuator spring. In the right picture, a proportionally countering return spring affects the actuator element.

Due to the different mechanical properties of a shape memory alloy in the high- and low-temperature phase, mechanical work can be performed while switching occurs between the two phases. The shear modulus can, for example, increase from $G_{LT} = 8.000$ N/mm to $G_{HT} = 24,000$ N/mm^2. In the low-temperature phase, the

S. Langbein (✉)
FG-INNOVATION GmbH, Universitätsstr. 142, 44799 Bochum, Germany
e-mail: langbein@gmx.com

K. Lygin
Ruhr-Universität Bochum, Universitätsstr. 150, 44801 Bochum, Germany
e-mail: konstantin.lygin@rub.de

© Springer International Publishing Switzerland 2015
A. Czechowicz, S. Langbein (eds.), *Shape Memory Alloy Valves*,
DOI 10.1007/978-3-319-19081-5_8

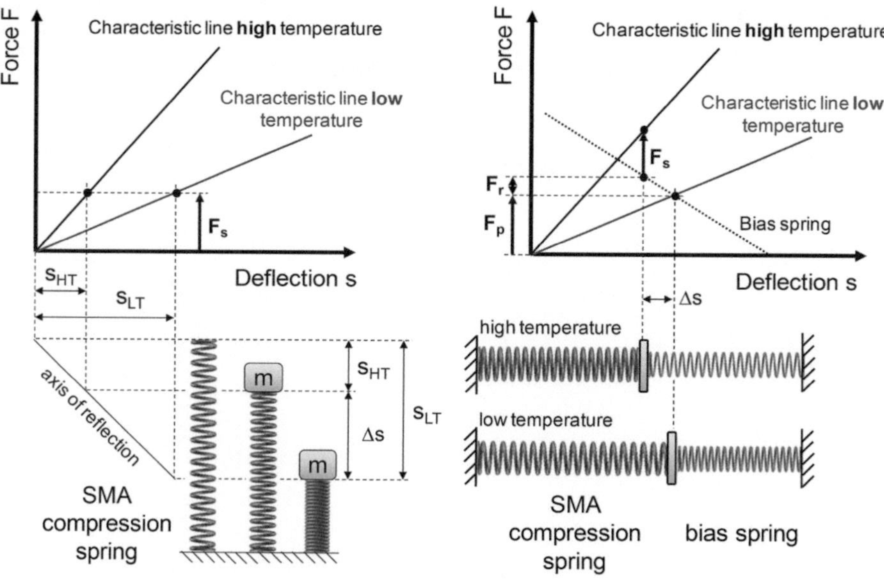

Fig. 8.1 Working principles of SMA-spring actuators

actuator is thus in the "soft" phase. Through the return element, in this case a constant mass or a bias spring, the actuator is deformed to the amount of s_{LT}. By increasing the temperature, the crystalline structure of the SMA material changes into the "hard" high-temperature phase. The actuator attempts to achieve its unstressed initial shape and develops a resistance force. In the high-temperature phase this force reaches its maximum value. The deformation to initial length is the amount of s_{HT}. The variance between deformations in the high- and the low-temperature phase is called working stroke Δs. Since the SMA spring is under mechanical load also in the high-temperature phase, it never reaches its unstressed initial length, as can be seen in the left graphic. F_s denotes the usable force for the control function. As indicated in Fig. 8.1, the usable actuator force resulting from a linearly increasing load (right graphic) is higher than the force resulting from a constant restoring mass (left graphic), because the gradient of the return spring characteristic requires balancing.

Usually, binary NiTi is used as actuator material for SMA actuators, however, ternary alloys, such as NiTiCu or NiTiFe, are used in technical applications as well. A key criterion for alloy selection is temperature hysteresis. Where a very low-temperature hysteresis of only a few Kelvin is required, NiTiFe or Ni-rich binary NiTi alloys are normally used. For average temperature hysteresis of about 10 K, NiTiCu alloys are suitable. If high-temperature hysteresis of approximately 20 K is acceptable, binary NiTi alloys can also be used. Coupled with the so-called thermo-mechanical treatment of semi-finished products, various actuator characteristics can be pre-set. The essential characteristics of an SMA-spring actuator are its conversion start- and finish-temperature (A_s and A_f temperature), the hysteresis width ΔT, and the shear moduli in the low- and high-temperature phase (G_{LT} and G_{HT}). In an ordinary

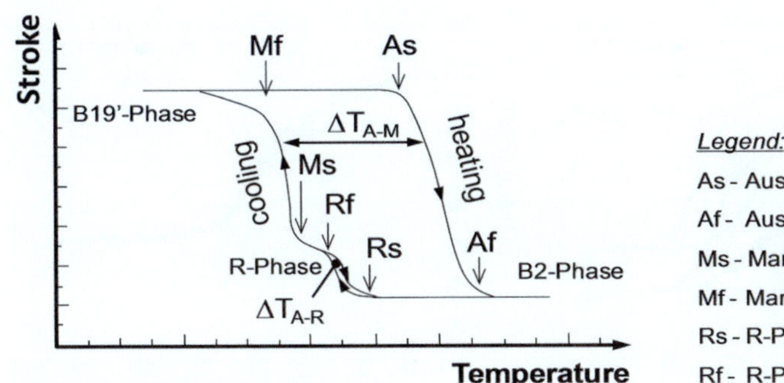

Fig. 8.2 Typical phase transformations in SMA

Table 8.1 Key data of transformations B2 > R and B2 > B19' according to [4]

Parameter of transformation from austenite (B2) to…	R-phase (B2 > R)	Martensite (B2 > B19')
SMA-effect size [%]	1	8
SMA-effect stability [−]	Very good	Good
Maximum transformation temperature [°C]	<50	<250
Width of hysteresis [°C]	<5	<80
Maximum stress [MPa]	370	900
Transformation enthalpy [J/g]	5	15–25
Durability, cycles [−]	$>10^7$	$\sim10^6$

transformation from austenite to martensite, austenite (A or B2, respectively) equals the high-temperature phase and martensite (M or B19', respectively) the low-temperature phase.

As described, both a very low and a very high-temperature hysteresis can be achieved with a binary NiTi alloy. The very low-temperature hysteresis is hereby achieved by utilizing an intermediate phase, the so-called R-phase, during which the transformation occurs from austenite to martensite. When the crystalline transformation is interrupted at an alloy-specific temperature during the R-phase, and the actuator is reheated, the very low-temperature hysteresis occurs (see Fig. 8.2). If, however, the actuator is cooled down from martensite to the R-phase and then reheated, the high-temperature hysteresis occurs. Both the transformation from austenite to R-phase (B2 > R) and the transformation from austenite to martensite (B2 > B19') offer great usability potentials for technical applications in Table 8.1.

Table 8.1 summarizes the relevant functional properties of both transformations. The transformation from austenite to R-phase provides a low achievable effect amount of max. 1 % strain for applications in actuator engineering [1]. The maximum effect amount of martensitic transformation from austenite to martensite offers, in comparison, up to 8 % strain [2]. In direct comparison with the martensitic transformation, the hysteresis of the R-phase transformation is very low with a few Kelvin only [3]. The martensitic transformation shows a considerably larger

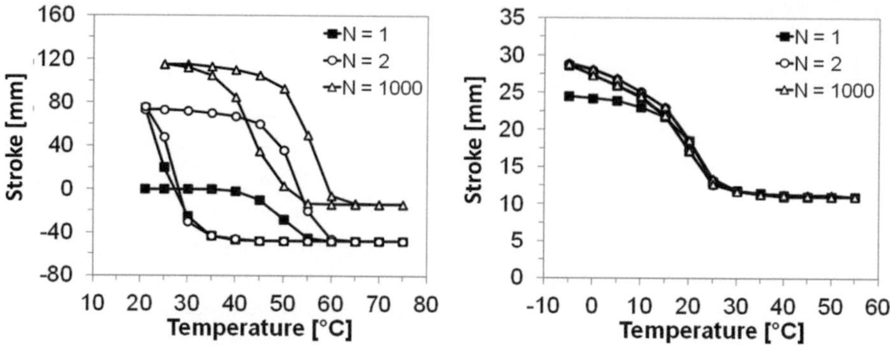

Fig. 8.3 Functional fatigue of a NiTi spring including transformation of austenite to martensite [6] and austenite to R-phase [7]

hysteresis of up to 80 °C [4]. Due to the smaller transformation enthalpy, the achievable working capacity of the R-phase transformation is only about 10–20 % of the martensitic transformation.

The use of the shape memory effect associated with the R-phase causes a very high number of cycles at almost constant effect amount [1]. In this comparison, the martensitic transformation presents some weaknesses with regard to effect stability and the achievable number of cycles [5]. Figure 8.3 shows the typical behavior of an SMA-spring actuator during thermal activation from martensite to austenite in the first 1,000 cycles. It is illustrated that (a) the positions of the martensite start and finish temperatures shift significantly, which leads to a reduction in temperature hysteresis, and (b) an elongation of the spring sample occurs and thus a displacement of the work area. The functional fatigue of NiTi-actuator alloys, however, happens primarily in the first few cycles. Through appropriate training measures, industrially produced actuator alloys are subjected to a planned fatigue during the production process, so that they show stable positioning behavior during the operation.

Usage of the R-phase transformation delivers the advantage, especially in actuator applications, that self-contained, temperature-sensitive and -controlled switching with small temperature hysteresis becomes adjustable. These product requirements prevail, amongst others, in applications used for heating and air conditioning technology, e.g. in thermostatic valves or jet nozzles.

8.2 Dimensioning of SMA Springs for Thermal SMA Valves

The following subsection supports geometric dimensioning of spring-based actuator systems after the extrinsic two-way effect, for example as may be applied in thermal proportional valves.

Based on the requirements of the actuator's regulation behavior, the mechanical properties of the used materials and the necessary constants, a guideline is presented. The guideline outlines, step by step, the design process of an SMA-spring actuator. Tension and compression spring actuators are discussed as operating principles of the SMA-spring actuator. Due to their geometry, these actuator principles possess mechanical transformation abilities between their displacement and positioning force. Thus a small amount of SM effect can be realized as a large control movement in a limited space, which would not be possible with ordinary SMA wire actuators. Figure 8.4 displays the guideline for dimensioning an SMA-spring actuator.

8.2.1   Step 1: Determination of Requirements

It starts with the definition (i.e. the provision of requirements) of the regulation behavior of the actuator system, which is expressed by the necessary positioning force F_S and the necessary displacement Δs.

8.2.2   Step 2: Determination of Material Properties

Determination of mechanical constants forms the second step of the design process. First, the shear moduli of the high-temperature phase G_{HT} and the low-temperature phase G_{LT} are defined. For SMA, the austenite phase equals the high-temperature phase. The low-temperature phase is, alloy actuator configuration-dependent, either the martensite or R-phase. The shear moduli can be determined from the tensile/ compression tests of the corresponding actuator configuration, which depends on parameters of their thermo-mechanical treatment. Common values for an austenite shear modulus range from 20,000 to 24,000 N/mm². Shear moduli of the low-temperature phase are usually between 6,000 and 8,000 N/mm².

Based on the approximately linear behavior of an SMA-spring actuator under mechanical stress, the shear modulus can be determined using the following simplified equation

$$G = \frac{\tau}{\gamma} = \frac{F}{s} \cdot \frac{8 \cdot D_m^{\,3} \cdot n}{d_w^{\,4}} \tag{8.1}$$

with:
G   Shear modulus
F   System force
s   Displacement at tensile test
τ   Shear stress
D_m Coil mean diameter
γ   Shear strain
d_w Wire diameter
n   Number of active windings

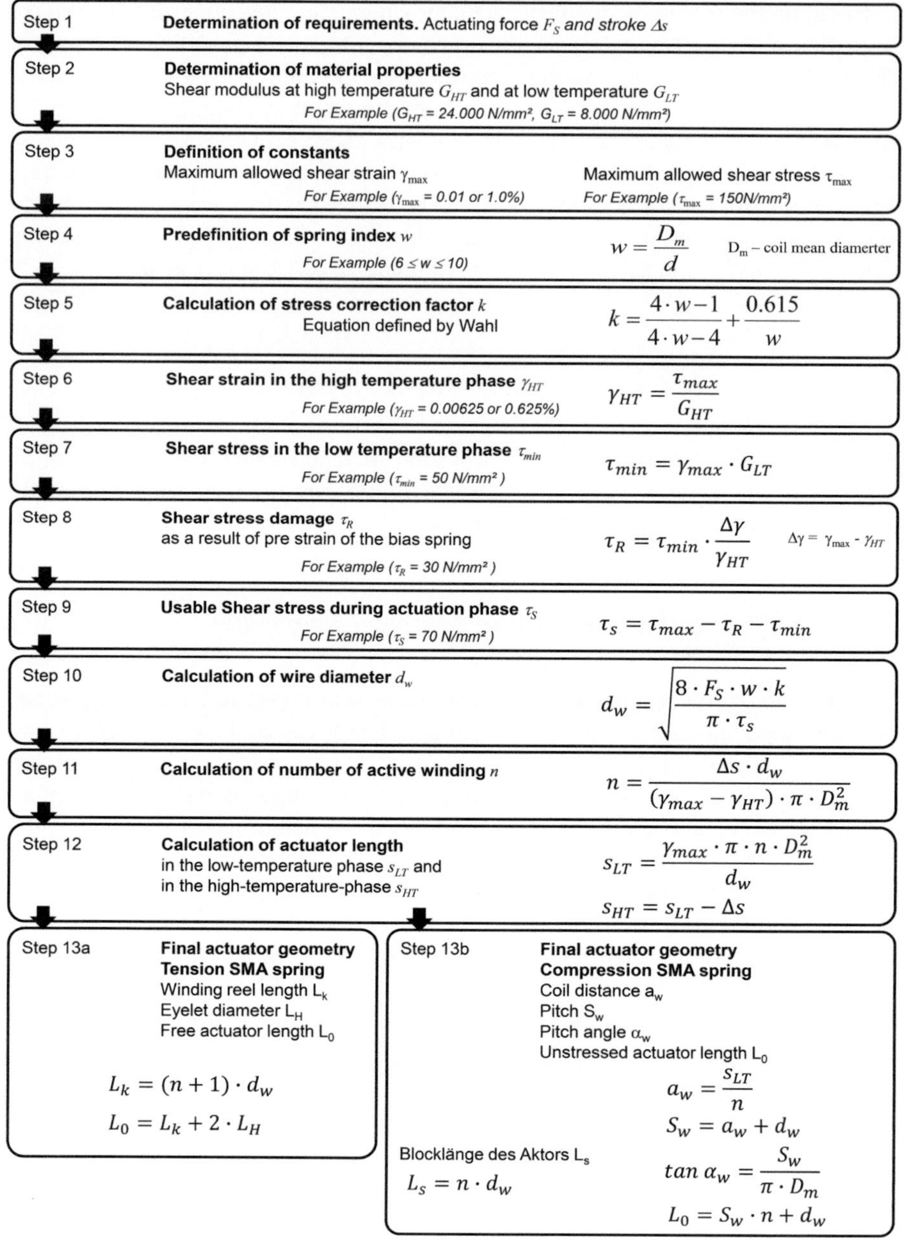

Fig. 8.4 Guideline for dimensioning of thermal SMA springs

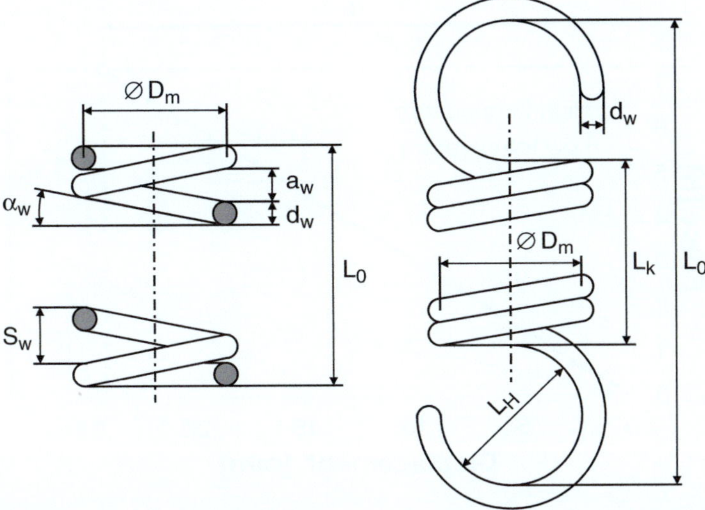

Fig. 8.5 Symbols used in spring design equations

Initially, no consideration was given to the stress correction factor k. This is further discussed in Step 5. The shear modulus is a geometry-independent quantity; therefore any SMA-spring geometry can be used for its determination, as long as they share the same alloy and thermo-mechanical treatment with the actuator that is being geometrically designed. The mechanical design parameters of a spring which are used in the design equations are presented in Fig. 8.5. Figure 8.6 shows a typical force–displacement diagram of an SMA tension spring at different temperatures. From this diagram, spring stiffness c and, consequently, shear modulus G can be determined.

8.2.3 Step 3: Definition of Constants

The third step defines values for the maximum permissible shear stress and the shear stress of the SMA-spring actuator. The limit for the maximum permissible shear stress of an SMA-spring actuator using the R-phase transformation should not, as shown by experience, exceed 1.0, in order to not endanger the low-temperature hysteresis [7]. Figure 8.7 shows the behavior of a tension spring actuator during the transformation from austenite to R-phase at varying shear strains between 0.6 and 1.8 %. It shows that the temperature hysteresis increases significantly at shear strains of over 1.0 %.

When the transformation from austenite to martensite is used, shear strains of 2 % or more can occur [1]. The shear strain can be derived from elongation of a spring actuator:

$$\gamma = \frac{s \cdot d_w}{\pi \cdot n \cdot D_m{}^2} \qquad (8.2)$$

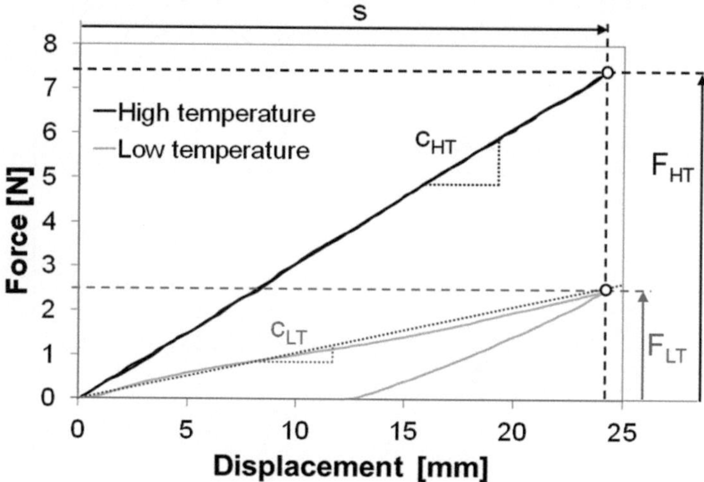

Fig. 8.6 Determination of shear modulus using the tensile test

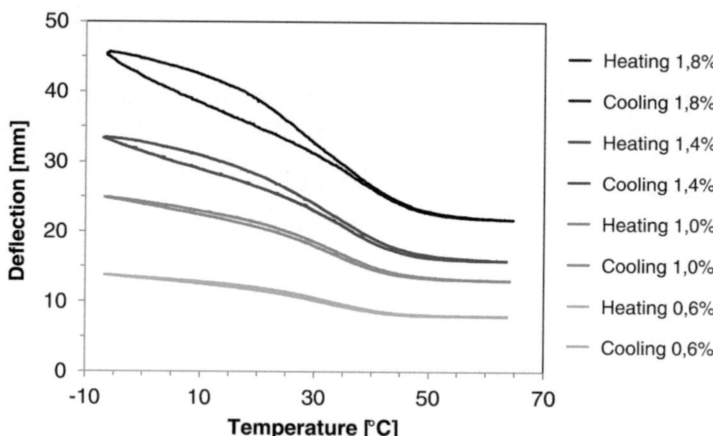

Fig. 8.7 Effects of different shear strains on SMA actuator behavior during the transformation from austenite to R-phase

The maximum permissible shear stress of an SMA-spring actuator may be chosen freely for R-phase transformations in the range between ca. 80 and ca.150 N/mm². For transformations into martensite, twice this value can be pre-set. Small permissible shear stresses cause high strokes, while low actuating forces and great shear stresses cause the opposite.

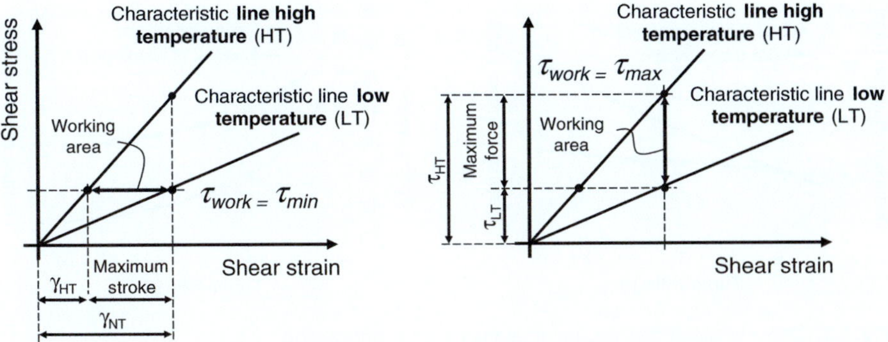

Fig. 8.8 Maximum stroke vs. maximum force of an SMA-spring actuator [7]

Shear stress can be derived from spring actuator force:

$$\tau = \frac{8 \cdot F \cdot D_m \cdot k}{\pi \cdot d_w^{\,3}} \qquad (8.3)$$

with: *k Stress correction factor*

Depending on the used value for maximum permissible shear stress, the actuator is either designed to provide a maximum force or a maximum displacement, as shown in Fig. 8.8. Choosing values in between causes a compromise between positioning force and displacement.

8.2.4 Step 4: Predefinition of Spring Index w

In the fourth step, the turns ratio w of the SMA-spring actuator is initially predefined to reduce the number of variables in the calculation. The turns ratio describes the quotient of the mean spring diameter D_m and the wire diameter d_w:

$$w = \frac{D_m}{d_w} \qquad (8.4)$$

For the design of SMA-spring actuators turns ratios between 6 and 10 are recommended. Experience shows that, from a manufacturing perspective, winding ratios of less than 6 are difficult to realize. However, they do offer advantages due to their high power output. Turns ratios of 10 or more are easier to manufacture, but they offer a lower power output. They are better suited for actuators with large displacement ranges and small actuating powers. Using the example of an SMA-spring

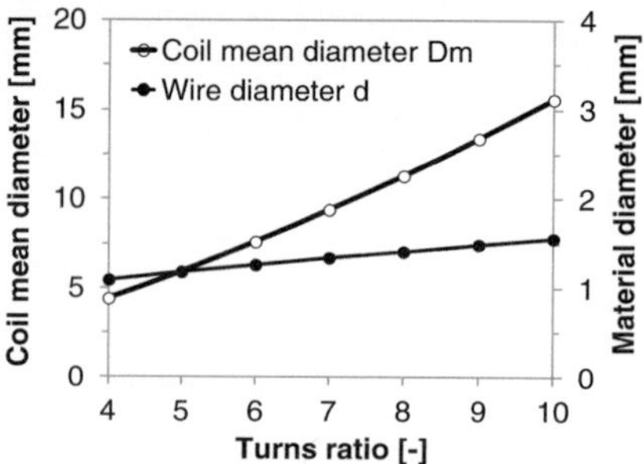
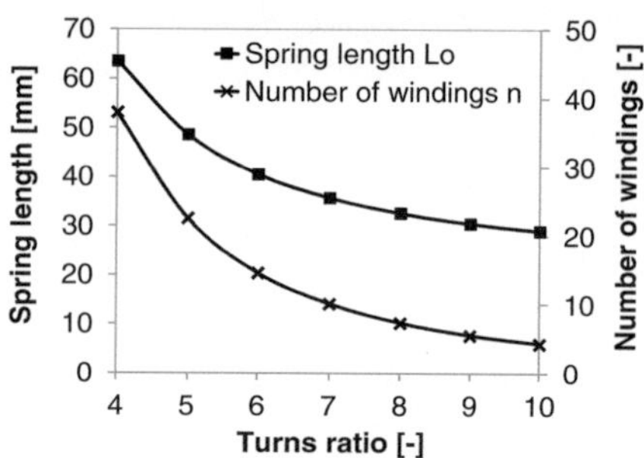

Fig. 8.9 Dependence of SMA actuator geometry on winding ratio

actuator with a required displacement range of $\Delta s = 10$ mm and a required force of $F_s = 10$ N, Fig. 8.9 shows geometrical variations dependent on the turns ratio w.

This comparison shows that low winding ratios cause low spring diameters, yet require high spring lengths. High winding ratios in turn cause high spring diameters, but also low spring lengths. Increasing the winding ratio from 4 to 10 causes a quadrupling of the spring diameter on the one hand, and on the other hand, halves the spring length. The wire diameter thereby increases by ca. 70 %. The required number of turns decreases by a factor of 9. The variation of the winding ratio thus offers great geometric design possibilities for spring actuators, which gives the developer the opportunity to optimally utilize the available space. Due to the regressive relationship between the values of winding ratio and spring length, or respectively, the number of spirals, the use of winding ratios over 6 is recommended, because value dependence decreases significantly from this value.

8.2.5   Step 5: Calculation of Stress Correction Factor k

The fifth step in actuator design is the determination of the stress correction factor k (using a equation of Wahl, see [8] or [9]), which considers inhomogeneous stress distribution in the wire cross section as a result of wire curvature. The stress correction factor is a purely geometry-dependent variable, which can be calculated with equations according to following authors: Bergsträsser [9], Göhner [8], Röver [10] or Wood [10]. However, the comparison of various calculation methods for factor k in the most common winding area between 6 and 10 showed no significant differences (Fig. 8.10). The selected stress correction factor k by Wahl can be determined using the following relationship according to Wahl:

$$k = \frac{4 \cdot w - 1}{4 \cdot w - 4} + \frac{0.615}{w} \tag{8.5}$$

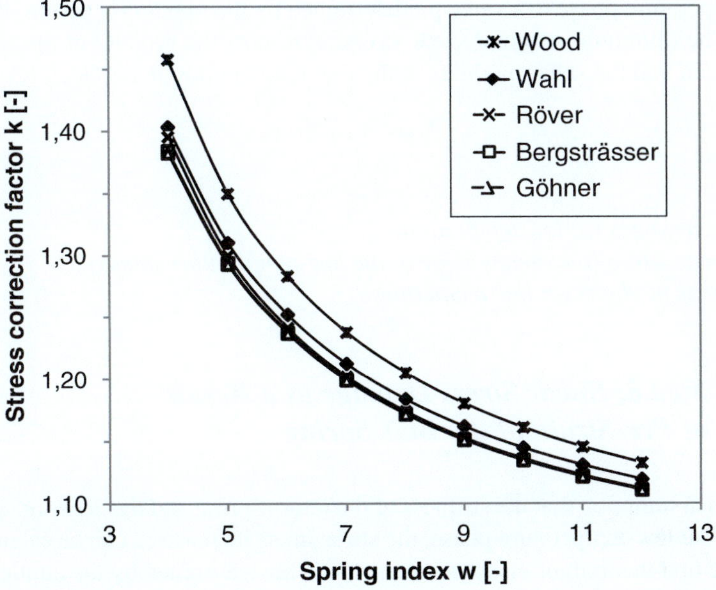

Fig. 8.10 Determination of stress correction factor k using various methods

8.2.6 Step 6: Shear Strain in the High-Temperature Phase

In the sixth design step, the shear strain (see Step 3) in the high-temperature phase is determined, which results from the ratio of the selected maximum permissible shear stress and the shear modulus in the high-temperature phase.

$$\gamma_{HT} = \frac{\tau_{max}}{G_{HT}} \qquad (8.6)$$

with:
γ_{HT} *Shear strain* (*at high temperature*)
τ_{max} *Shear stress* (*as maximum defined shear stress*)
G_{HT} *Shear modulus* (*high temperature*)

8.2.7 Step 7: Shear Stress in the Low-Temperature Phase

The shear stress τ_{min} is the minimum load on the actuator in the low-temperature phase. The pre-application of this shear stress is required in order for the actuator to function normally under to the two-way effect. The actuator is thus deformed in the low-temperature phase to a value predetermined by the maximum permissible shear strain. The minimum required shear stress represents the product of the maximum shear strain and the shear modulus in the low-temperature phase G_{LT}.

$$\tau_{min} = \gamma_{max} \cdot G_{LT} \tag{8.7}$$

with:

τ_{min} *Shear stress (at low temperature)*
γ_{max} *Shear strain (maximum value in the low-temperature phase)*
G_{LT} *Shear modulus (at low temperature)*

8.2.8   Step 8: Shear Stress Damage as a Result of Pre-Strain of the Bias Spring

Based on assumption that the stiffness of the return spring and the actuator spring are equal in the low-temperature phase, the shear stress increase τ_R can be calculated by using the first theorem of incepting lines. It can be determined by the quotient of the usable shear strain $\Delta\gamma$ and the maximum shear strain γ_{max} in the low-temperature phase multiplied by the shear stress in the low-temperature phase τ_{min}

$$\tau_R = \tau_{min} \cdot \frac{\Delta\gamma}{\gamma_{HT}} \tag{8.8}$$

with:

τ_R *Shear stress as a result of pre-strain of the bias spring*
τ_{min} *Shear stress (value at low temperature)*
$\Delta\gamma$ *For the actuation usable shear strain ($\gamma_{max} - \gamma_{HT}$)*
γ_{HT} *Shear strain at high temperature*

8.2.9   Step 9: Usable Shear Stress During Actuation Phase

The usable shear stress τ_S is composed of the difference of the maximum permissible shear stress τ_{max}, the increase in shear stress caused by the return movement of the bias spring τ_R, and the minimum shear stress required for resetting τ_{min}. Since the required values were defined previously, the usable shear stress τ_S can now be calculated.

$$\tau_s = \tau_{max} - \tau_R - \tau_{min} \tag{8.9}$$

with:

τ_s *Usable shear stress*
τ_{max} *Shear stress (as maximum defined shear stress)*
τ_R *Shear stress as a result of pre-strain of the bias spring*
τ_{min} *Shear stress (value at low temperature)*

8.2.10 Step 10: Calculation of Wire Diameter

The following step focuses on the wire diameter d_w of the spring actuator taking into account the available usable shear stress τ_S, the required positioning force F_S and the selected coil diameter w. After determining the wire diameter d_w, the mean coil diameter D_m can be calculated with the aid of the winding ratio w.

$$d_w = \sqrt{\frac{8 \cdot F_S \cdot w \cdot k}{\pi \cdot \tau_s}}$$

(8.10)

with:

d_w	*Wire diameter*
F_s	*Operating Force*
w	*Winding ration*
k	*Stress correction factor*
n	*Number of active windings*
τ_S	*Usable shear stress*

8.2.11 Step 11: Calculation of Number of Active Winding

Finally, only the number of required active actuator windings remains to be determined, taking into account the required displacement Δs and the permissible strain of the spring $\gamma_{\max}$ and γ_{HT}.

$$n = \frac{\Delta s \cdot d_w}{\left(\gamma_{\max} - \gamma_{HT}\right) \cdot \pi \cdot D_m^2}$$

(8.11)

8.2.12 Step 11: Calculation of Actuator Length in the High- and in the Low-Temperature Phase

The following two sub-steps cover the actuating range in the low- and high-temperature phase s_{LT} or s_{HT}, respectively. Knowing these two values is of particular interest with regards to the spring actuator's space requirements.

$$s_{LT} = \frac{\gamma_{\max} \cdot \pi \cdot n \cdot D_m^2}{d_w}$$

(8.12)

with: s_{LT} *actuator length in the low temperature*

$$s_{HT} = s_{LT} - \Delta s$$

(8.13)

with:

s_{HT}	*Actuator length in the high temperature*
s_{LT}	*Actuator length in the low temperature*
Δs	*Usable stroke*

8.2.13　Step 12: Final Actuator Geometry

The following steps cover the finalization of actuator geometry. It is dependent on the selected spring form, whereby tension springs and compression springs are the most commonly chosen geometrical forms.

If a tension spring actuator is preferred, then the winding reel length L_k and the total spring length L_0 are determined according to DIN EN 13906-2 [9] in the last two steps. Since all windings of a tension spring are usually joined, the length of the winding reel L_k results from the number of windings $n+1$ multiplied by the wire diameter d_w. The spring length L_0 is composed of the winding reel length L_k and twice the height of the spring eyelet L_H.

$$L_k = (n+1) \cdot d_w \tag{8.14}$$

with:

L_k	*Actuator winding reel length*
n	*Number of active winding*
d_w	*Wire diameter*

$$L_0 = L_k + 2 \cdot L_H \tag{8.15}$$

with:

L_0	*Free actuator length*
L_k	*Actuator winding reel length*
L_H	*Eyelet diameter*

If a compression spring is the preferred actuator form, the design layout is a bit more complex, because the spring—in unloaded condition—must have a precise coil distance in order to ensure full functionality in operation. Calculation is done in accordance with the german standard DIN EN 13906-1 [11]. First, the block length of the actuator L_s is calculated to determine the minimum space requirement.

$$L_s = n \cdot d_w \tag{8.16}$$

with:

L_s	*Actuator spring solid length*
n	*Number of active winding*
d_w	*Wire diameter*

Then the required coil distance a_w is determined, which results from the maximum possible spring elongation in the low-temperature phase s_{LT}. When compressing the spring under load, the actuator is thus not loaded beyond the permissible value of the shear strain γ_{LT}, because it always comes to stall in the block position.

$$a_w = \frac{s_{LT}}{n} \tag{8.17}$$

with:

a_w *Coil distance*
s_{LT} *Actuator length in the low-temperature phase*
n *Number of active winding*

When the coil diameter a_w is subsequently added to the wire diameter d_w, we get the winding pitch S_w.

$$S_w = a_w + d_w \tag{8.18}$$

with:

S_w *Pitch*
a_w *Coil distance*
d_w *Wire diameter*

This value can now be used to determine the pitch angle of the compression spring α_w

$$\tan \alpha_w = \frac{S_w}{\pi \cdot D_m} \tag{8.19}$$

Finally, the spring length is calculated in its unstressed state L_0.

$$L_0 = S_w \cdot n + d_w \tag{8.20}$$

The design of the SMA compression spring is hereby complete. The calculation of the return spring is subject to the constraint that the spring rate c of the return spring should have the same value as the SSMA spring in the low temperature phase, in accordance with DIN EN 13906.

References

1. I. Ohkata, Y. Suzuki, The design of SMA actuators and their applications, in *Shape Memory Materials*, ed. by K. Otsuka, C.M. Wayman (Cambridge University Press, Cambridge, 1998)
2. D.C. Lagoudas (ed.), *Shape Memory Alloys. Modeling and Engineering Applications* (Springer, New York, NY, 2008)
3. Y. Zhoua, G. Fana, J. Zhang, X. Ding, X. Ren, J. Suna, K. Otsuka, Understanding of multistage R-phase transformation in aged Ni-rich Ti–Ni shape memory alloys. Mater. Sci. Eng. A **438–440**, 602–607 (2006)

4. M. Mertmann, *NiTi-Formgedächtnislegierungen für Aktoren der Greiftechnik* (VDI Verlag, Düsseldorf, 1997)
5. M. Mertmann, in *Fatigue in NiTinol Actuators*. Conference proceedings of Actuator 06, 10th International Conference on New Actuators (Bremen, 2006), pp. 461–466
6. G. Eggeler, E. Hornbogen, A. Yawny, A. Heckmann, M. Wagner, Structural and functional fatigue of NiTi shape memory alloys. Mat. Sci. Eng. A **378**, 24–33 (2004). Elsevier
7. K. Lygin, *Eine Methodik zur Entwicklung von umgebungsaktivierten FG-Aktoren mit geringer thermischer Hysterese* (Dr. Hut Verlag, München, 2014)
8. M. Meissner, H.-J. Schorcht, *Metallfedern. Grundlagen, Werkstoffe, Berechnung, Gestaltung und Rechnereinsatz. 2. Auflage* (Springer Verlag, Berlin, 2007)
9. EN 13906-2, Cylindrical helical springs made from round wire and bar. Calculation and design Part 2: Extension springs (Beuth Verlag, Berlin, 2002)
10. H. Funakubo: *Shape Memory Alloys*. CRC Press, New York 1987
11. EN 13906-1, *Cylindrical helical springs made from round wire and bar. Calculation and design Part 1: Compression springs* (Beuth Verlag, Berlin, 2002)

Chapter 9
Design of Electrical SMA Valves

Alexander Czechowicz

9.1 Electrical SMA Actuators: Fundamental Effects and System Design

Electrical SMA actuators mainly use the intrinsic heating function by Joule's law to convert electrical energy into thermal energy. The SMA actuator itself converts the thermal energy into mechanical work output, and thermal leaks in the form of convection, thermal radiation and thermal conduction. In this chapter, the design of electrical SMA actuators is discussed mainly related to SMA wire actuators. SMA wires, especially when used as longitudinal actuators, can generate relative high pulling forces in comparison to their cross section. Due to their small diameters, they are suitable for direct electrical heating. At increasing diameter, the absolute resistance level of the SMA wire decreases which cause a need of very high current levels for direct electrical activation.

Within this chapter, first the basic working principles are discussed and then a mechanical fast-track calculation of the main design parameters. Afterwards, activation principles are presented as well as examples of electrified activations of SMA wire actuators. A numerical model for the balance equation, as shown in Fig. 9.1, shows the possibility to find exact solutions for specific designs. The chapter is finalized by an exemplary development of a squash-valve using SMA wires.

A. Czechowicz (✉)
Zentrum für Angewandte Formgedächtnistechnik, Forschungsgemeinschaft Werkzeuge und Werkstoffe e.V., Papenberger Straße 49, 42859 Remscheid, Germany
e-mail: czechowicz@fgw.de

© Springer International Publishing Switzerland 2015
A. Czechowicz, S. Langbein (eds.), *Shape Memory Alloy Valves*,
DOI 10.1007/978-3-319-19081-5_9

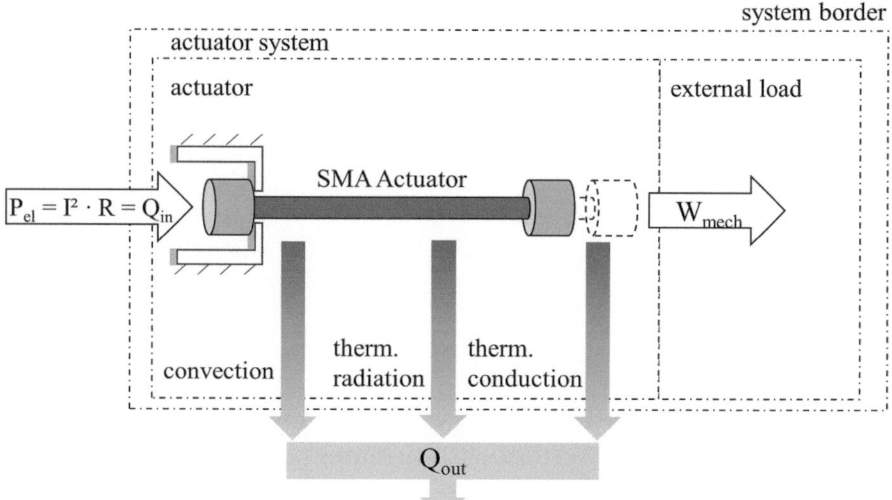

Fig. 9.1 Basic understanding of electrical SMA straight wire actuator

9.2 Hindrances During the Development of SMA Valve Drives

The behavior of electric SMA actuators is far more complex than in thermal applications. This fact is the main hindrance for the assertion of SMA on the market. Due to thermal balances, mechanical behavior and a multitudinous effect-network, a simple-looking shape memory wire is a complicated high-tech system. This is often disregarded by inexperienced engineers who try to develop SMA valve drives. In comparison to conventional drive principles like solenoids, the dynamic behavior depends on a great number of critical parameters. Figure 9.2 shows exemplary the influence network of SMA which has to be passed during the design process of electrical SMA valves against state-of-the-art systems.

If an unexperienced engineer has to change several parameters in order to optimize the functionality, a complete redesign of the system is necessary. Additionally, neither factors like the dynamic characteristic nor the fatigue behavior are listed in the diagram. Therefore, experimental approaches to the problem are necessary. With this, the development costs and time increase, so the developing engineers often fall back to familiar actuator systems.

To date, there are several ideas how to manage the challenges in the development of SMA actuators. One solution in order to simplify the development process is to standardize the technology itself. With the usage of standardized SMA elements, the peripheral equipment and testing methods, shape memory valve drives can be developed faster [2].

Another way to achieve an effective engineering process is to standardize the development of SMA actuators itself. Therefore, a stepped method with differently

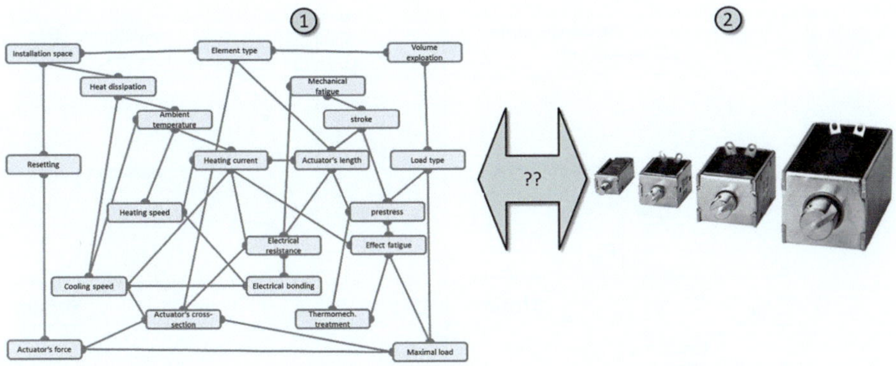

Fig. 9.2 Influence network of a SMA actuator system in comparison to state-of-the-art solutions [1]

focused levels has to be used to clarify the details of a system. This will be discussed in Chap. 10, presenting a development method for SMA actuators. But still such ideas do not clarify the multitudinous characteristics of SMA elements to the system developer.

As for simple systems, a fast-track development can be used for first simple design parameters of straight SMA wires, which will be focused in this chapter. For a more detailed calculation, a simulation tool for SMA elements is needed which has to fulfill certain requirements:

1. The simulation handling has to be simple.
2. Due to frequent systems' optimizations, the calculation time has to be very fast.
3. An intelligent supporting system has varied the systems' parameters in order to achieve the best actuators' performance.

These needs can be fulfilled by different approaches. Several attempts have been already made to create simulation tools, one of these ideas is presented in the following chapter.

9.3 SMA Wires as Electrical Actuators

To visualize the behavior of SMA wires, a simplification in mind can be done by determination of the cold and hot state of SMA wires as independent stress–strain, or force–strain, characteristic. This is shown in Fig. 9.3 by the plateau curves in martensitic and austenitic state of the SMA wire. The left part of the figure describes the SMA actuation against a constant force, represented by a constant mass. The usable force is a difference between the force affected by the mass (F_m) and the maximal reachable stress level in austenitic state. These factors are related to the reachable strains. As shown in the diagram, a complete retransformation is not possible at loaded SMA wire. There is a residual strain (ε_h) which is related to the force–strain characteristic of the SMA wire in austenitic state. Note that at constant mass,

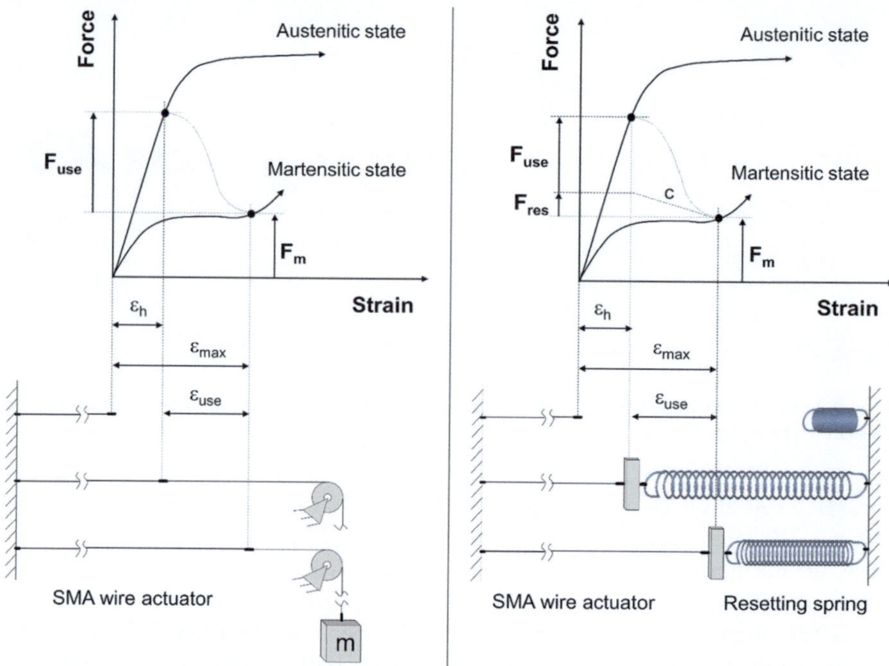

Fig. 9.3 SMA wire actuators working against constant mass (*left*) and resetting spring (*right*) [1]

the force in the system would stay constant. But the increase of the force resembles its power potential.

The right side of Fig. 9.3 presents the actuation of SMA wire against a conventional spring. In addition to the force potential, the real system force is determined by the usable force and the force which is needed to retransform the contracted SMA wire into the elongated element. Note that the resetting stress in general can be chosen between 20 and 50 N/mm² relating to different SMA alloys during the cooling from austenitic to martensitic state. The elongation stress in pure martensitic state is significantly higher (40–80 N/mm²). This is caused by undirected phase elements in the SMA wire which act in resistance against the deformation.

As basic configuration the SMA wire can be used for longitudinal actuation. Normally, one end of the SMA wire is mounted fixed to the housing, while the other end is connected to a movable slider and the resetting element.

The stroke of the straight SMA wire actuator can be calculated to:

$$s_{use} = l_0 \cdot \varepsilon \tag{9.1}$$

As well as the force can be calculated by the following equation:

$$F_{min} = \sigma_{min} \cdot A_w \tag{9.2}$$

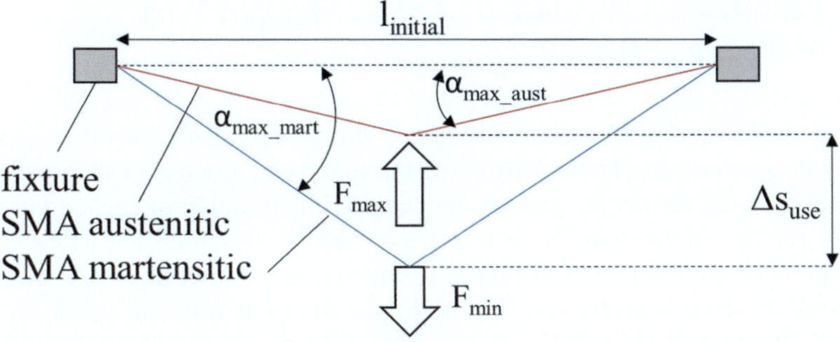

Fig. 9.4 Alternate SMA wire configuration in form of a bow

Whereas A_w is the SMA wire's cross section area.

An alternate configuration can also be achieved by fixing both ends of the SMA wire and connecting a slider with a resetting element in the midsection of the wire. Such "bow"-like configurations have been often used as an internal gearing for stroke. Figure 9.4 shows such a configuration with simplified calculation parameters in order to calculate the system's translation.

The determination of the critical working angles α_{max_aust} and α_{max_mart} are vital for the maximal stroke and output force levels. With increasing working angles the stroke decreases while the maximal output force increases. At $\alpha \approx 90°$, the stroke is equal to a straight wire of $l_0/2$ length. Due to the doubled cross-section (equal to two longitudinal SMA wires), the maximal force is doubled in this case.

The maximal stroke can be calculated by the utilization of the trigonometric calculation:

$$s_{use} = \sin\left[\arccos\left(\frac{1}{1+\delta}\right)\right] \cdot \frac{\frac{l_0}{2} \cdot (1+\delta)}{2} \tag{9.3}$$

with δ = contraction of the SMA wire.

Regarding the transmission factor of the usable stroke in relation to the generated longitudinal contraction of the SMA wire, the translation can be described as:

$$k = \frac{s_{use}}{l_0 \cdot \delta} \tag{9.4}$$

For simplified calculation of the minimal force which is necessary for the resetting of the SMA wire, the geometrical relation is given by:

$$F_{min} = \sigma_{min} \cdot \frac{\pi}{2} \cdot d^2 \cdot \sin\left[\arccos\left(\frac{1}{(1+\delta)}\right)\right] \tag{9.5}$$

9.4 Fast-Track Calculation of SMA Straight Wire Mechanical Design

Starting with a requirement determination, a simplified calculation of the electrical SMA wire actuators can be done by going through the diagram presented in Fig. 9.5. The usable force, the stroke, and the needed lifetime (number of cycles) have to be defined for the application. Note that these factors are connected to each other, therefore several combinations of these parameters are not possible. Afterward, it is necessary to select a stress level range in combination with the strain level and lifetime. Diagrams such as in Fig. 9.5 contain all necessary data for such selection.

Note that the presented diagram is made for a specific SMA wire product, and is not necessary applicable on every SMA element/wire. To be safe, a complete product development should contain an experimental verification of the minimal and maximal applicable stress–strain factors concerning the system design as shown in Fig. 9.5.

In the next step, the usable stress can be calculated just by subtraction of the minimal stress from the maximal stress level. Afterward, it is necessary to determine the resetting of the SMA wire. If a constant mass is used, the diameter of a round SMA wire can be obtained directly from the equation for d_w in Fig. 9.5. The resetting mass is a function of the minimal stress and the wire diameter.

For linear resetting element (spring element), it is necessary to determine the spring constant c first. Due to the connection between the SMA wire's stroke and

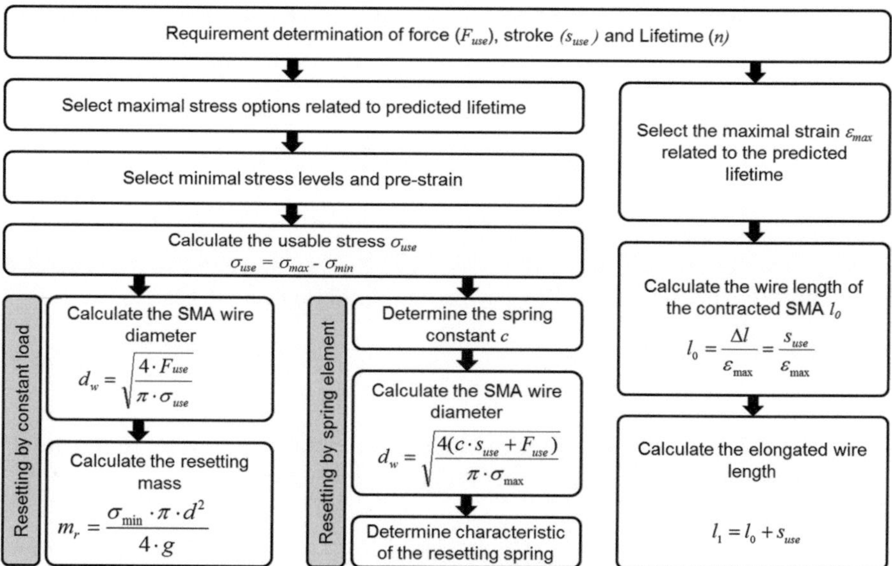

Fig. 9.5 Fast-track calculation of mechanical SMA straight wire design

Table 9.1 Solutions for different maximal stress levels obtained by fast-track calculation

F_{min} [N]	F_{max} [N]	c [N/mm]	s_{use} [mm]	σ_{max} [N/mm^2]	d_w [mm]	σ_{min} [N/mm^2]	σ_{use} [N/mm^2]
10	14	1.6	2.5	50	0.6	35.39	14.61
10	14	1.6	2.5	100	0.42	72.22	27.78
10	14	1.6	2.5	150	0.34	110.2	39.8
10	14	1.6	2.5	200	0.3	141.54	58.46
10	14	1.6	2.5	250	0.27	174.74	75.26
10	14	1.6	2.5	300	0.24	221.16	78.84
10	14	1.6	2.5	350	0.23	240.81	109.19
10	14	1.6	2.5	400	0.21	288.86	111.14

the total force, the SMA wire's diameter is a function of the SMA wire's usable stroke. For simplification, the usable stroke can be seen as maximal stroke in the calculation of the diameter.

The right path of the diagram in Fig. 9.5 shows the calculation of the elongated SMA wire. The strain level, selected also from the diagram in Fig. 9.5, can be inserted in the equation for the calculation of l_1. The maximal strain, multiplied with the initial length, results in the usable stroke.

For a fast-track calculation a simple example can be discussed here:

If an SMA actuator for valve piston movement is assumed with a displacement of 2.5 mm and a resetting force level of 10 N (closed) and 14 N (opened), a maximal stress level up to 400 N/mm^2 should be used. From $F_{min} = 10$ N and $F_{max} = 14$ N the used spring constant c can be determined as $c = 1.6$ N/mm.

Table 9.1 presents an overview for this problem calculated for different maximal stress levels from 50 to 400 N/mm^2. By the usage of the equation for resetting by spring element (Fig. 9.5), the diameter d_w can be calculated.

The minimal stress level, also used as pre-stress of the system, can be obtained from

$$\sigma_{min} = \frac{F_{min}}{\dfrac{d_w^2}{4} \cdot \pi} \tag{9.6}$$

If the SMA wire works against a constant load, the minimal and maximal stress levels are equal. In the next step of the fast-track calculation, the needed contracted SMA wire length can be obtained by usage of the right path in Fig. 6.4 with Eq. (9.1) transformed to:

$$\frac{s_{use}}{\varepsilon} = l_0 \tag{9.7}$$

In Table 9.2, different contracted wire lengths have been calculated for different strain levels. The pre-strained SMA wire can be calculated by:

$$l_1 = l_0 + s_{use} \tag{9.8}$$

Table 9.2 Solutions for different elongations obtained by fast-track calculation

s_{use} [mm]	ε_0 [%]	ε [ε_0/100]	l_0 [mm]	l_1 [mm]
2.5	0.5	0.005	500	502.5
2.5	1	0.01	250	252.5
2.5	1.5	0.015	166.7	169.2
2.5	2	0.02	125	127.5
2.5	2.5	0.025	100	102.5
2.5	3	0.03	83.3	85.8
2.5	3.5	0.035	71.4	73.9

9.5　Numerical Simulation of SMA Wire Actuators

Getting an overview of the SMA element's characteristics is a challenging, but not an impossible, task. To build a simulation, it is necessary to determine the simulation's simplification which will satisfy the user's needs. In terms of reducing the shape memory effect to a mathematical equation, Fig. 9.3 shows the actuator system's thermal balance. The system consists of a simplified two-dimensional actuator itself and the mechanical (external) load. The thermal and mechanical energies have to be put on a level with the electrical energy during the heating phase. This is shown in Eq. (9.1). The stored thermal energy, convection, thermal radiation, the latent heat for transformation and the mechanical energy, as well as the electrical energy are elements which are contained in the parameter network discussed before. The activation of SMA elements, defined as the stroke–time relation of the actuator within transformation is solely specified by the velocity of cooling and heating the material. Especially with time variable current profiles and also variable load conditions, these dynamic parameters cannot or only inadequately be considered with simplifying algebraic functions.

According to calculations for a straight SMA-wire, heat transmission caused by free air convection is about 90 % and by heat radiation is about 10 % of the total emission. The low influence of heat conduction due to the clamping can therefore be neglected a priori. Similarly to the consideration of the emitted heat flow, the required mechanical energy, resulting from the returning device and external loads as well as the integral latent heat for transformation from martensite to austenite has to be quantified. The required latent heat for transformation is proportional to the derivation of the volumetric martensite fraction ξ [3].

$$\underbrace{c_p \cdot \rho \cdot V_W \cdot \frac{\mathrm{d}T_W}{\mathrm{d}t}}_{\text{stored thermal energy}} + \underbrace{\alpha \cdot F_W \cdot (T_W - T_\infty)}_{\text{convection}} + \underbrace{\varepsilon_m \cdot \sigma_{\text{rad}} \cdot F_W \cdot \left(T_W^4 - T_\infty^4\right)}_{\text{thermal radiation}}$$

$$+ \underbrace{\rho \cdot V_W \cdot \Delta H \cdot \left|\frac{\mathrm{d}\xi}{\mathrm{d}t}\right|}_{\text{latent heat for transformation}} + \underbrace{\frac{\mathrm{d}W_{\text{mech}}}{\mathrm{d}t}}_{\text{mech. energy}} = \underbrace{I^2\left(t\right) \cdot R}_{\text{electrical energy}} \qquad (9.9)$$

Colling: $I^2\left(t\right) \cdot R_D = 0$

with:	c_P:	Specific heat capacity	V_W:	Actuator volume
	ε_m:	Emission ratio	F_W:	Actuator surface
	σ_{rad}:	Stefan–Boltzmann constant	T_W:	Wire temperature
	ΔH:	Integr. latent heat for transformation	T_∞:	Ambient temperature
	ξ:	Volumetric martensite fraction	W_{mech}:	Mechanical energy
	ρ:	Density	α	Conv. heat transfer coefficient

Equation (9.1) is a nonlinear first-order differential equation which cannot be solved definitely with analytic methods. Thus, a solution curve is calculated numerically by using an approximation procedure within the MATLAB/ SIMULINK environment. The main calculation parameters are:

- *Convective Heat-Transfer Coefficient α:*
  The heat transfer coefficient α is the characteristic parameter for convective heat transfer depending on the wire temperature and diameter.

$$\alpha\left(T_W, d_W\right)$$

  The numerical model is based on a polynomial approximation of α for each examined wire diameter.
- *The wire's ohmic resistance* corresponds to the ratio of martensite (ρ_M) and austenite (ρ_A) during phase transformation and to the current actuator's geometry. Further explanations of the resistance characteristic of SMA are given in Chap. 5.
- *Geometry*: When contracting, a variation of the SMA wire's geometry values (length, diameter, surface) is induced. Provided that the actuator's volume is constant, these values are determined in real-time during simulation and linked to the calculation of other process parameters (e.g. the ohmic resistance).
- *Thermal expansion ε-therm*: Besides the contraction of the actuator in consequence of the shape memory effect, a contrary thermal expansion of the actuator has to be considered. The different coefficients of thermal expansion of martensite and austenite (β_M, β_A) have to be taken into account.

$$\varepsilon_{\text{therm}} = \left(T_W - T_\infty\right) \cdot \left(\xi \cdot \beta_M + \left(1 - \xi\right) \cdot \beta_A\right) \tag{9.10}$$

- *Elastic elongation ε_{elast}*: The conventional elastic elongation $\varepsilon_{\text{elast}}$ of the wire actuator depends, apart from the actuator geometry and of course the stress, basically on the actuator's structure. The different Young's moduli of martensite and austenite and the phase mixture during transformation have to be considered [4].

$$\varepsilon_{\text{elast}} = \sigma \cdot \left(\frac{\xi}{E_M} + \frac{1 - \xi}{E_A}\right) \tag{9.11}$$

- *Influence of stress on the transformation temperatures*: The mechanical stress state inside the wire affects the value of the transformation temperatures.

According to the general characteristic of SMA presented in Chap. 3, the transformation temperatures are in linear proportion to the actuator's mechanical stress condition. The phase transformation from martensite to austenite and vice versa is therefore not only thermal- but also stress-induced, as the actuator's stress condition is not constant due to the use of bias springs. Thus M_f, M_s, A_s, and A_f need to be iteratively adjusted in the numerical model to account for the current mechanical stress in the actuator wire.

For the analytic description of the SMA's characteristics during the phase transformation, an interconnection with a hysteresis model that has to be parameterized by the input of specific actuator attributes is necessary. For this purpose, a simple hysteresis model using mathematical descriptions is used. The tanh model [5], describing the transformation hysteresis as a trigonometric function of the transformation and wire temperatures, is shown in Eqs. (9.12) and (9.13):

$$\xi\left(T_d\right) = \frac{1-\xi^\circ}{2}\tanh\left[2\left(\frac{M_f-T_d}{M_s-M_f}+\frac{1}{2}\right)\right]+\frac{1+\xi^\circ}{2} \quad \text{for } \xi^\circ > 0 \qquad (9.12)$$

$$\xi\left(T_d\right) = \frac{\xi^\circ}{2}\tanh\left[2\left(\frac{A_s-T_d}{A_f-A_s}+\frac{1}{2}\right)\right]+\frac{\xi^\circ}{2} \quad \text{for } \xi^\circ < 0 \qquad (9.13)$$

with:

A_s=*Austenite start temperature*
A_f=*Austenite finish temperature*
M_s=*Martensite start temperature*
M_f=*Martensite finish temperature*
T_d=*SMA-wire temperature*

The numerical model implemented in MATLAB/SIMULINK considers all these relevant parameters in order to simulate the time response of different SMA elements (wire, spring, and ribbon material) during phase transformation.

The following figures show the results of the simulation compared to the experimental results. This comparison confirms the physically good implementation of the actuator's properties and its behavior in the SIMULINK®-model exemplary (Fig. 9.6).

9.6 Application Characteristics of Electric SMA Actuator Systems

9.6.1 Operating Temperatures

Hence the function of electrical SMA actuators still depends on the thermal-induced phase transformation, the link to ambient temperatures, besides the dynamic problems as discussed in Chap. 3, has a major impact on the system's functionality

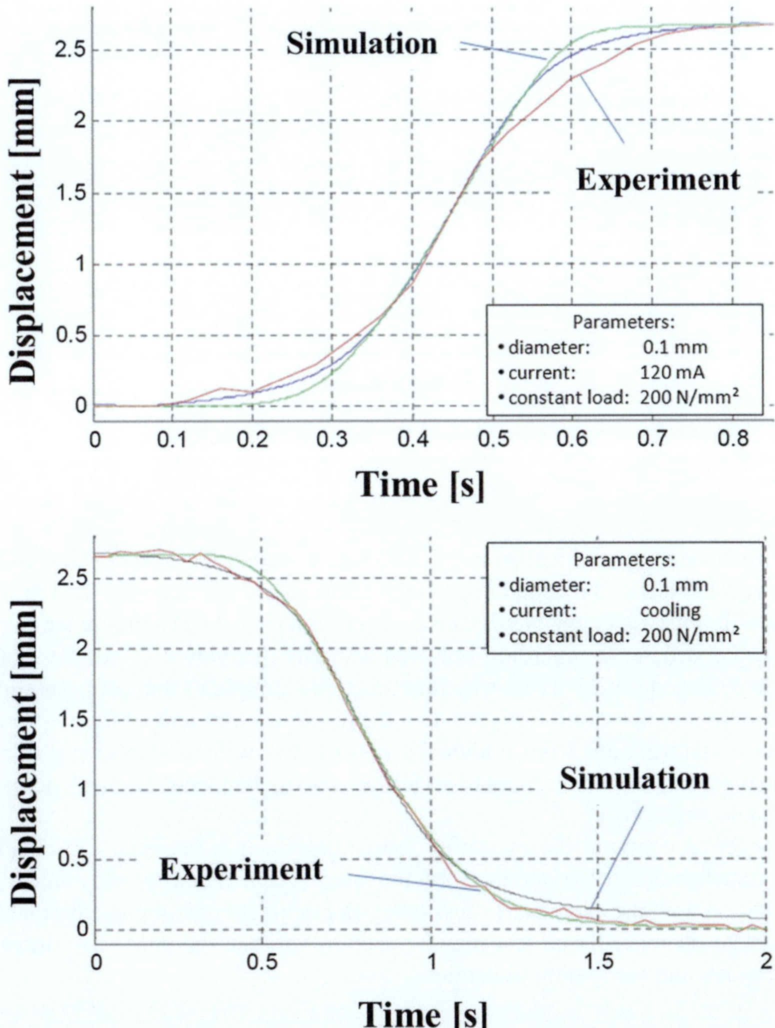

Fig. 9.6 Comparison between the numerical simulation and experimental validation (SMA-wire: diameter 0.1 mm, length 90 mm, load 200 N/mm², heating time 0.8 s) [6]

and reliability. Figure 9.7 shows an exemplary hysteresis of an electrical SMA actuator in different ambient temperature fields. As discussed above, the temperature of the SMA actuator is induced by the relation of the electrical input power to the power output dissipation. If the ambient temperature is lower than the lowest phase transformation temperature (martensite finish temperature), a full functionality of the system, disregarding the transformation dynamics, is possible. If the ambient temperature is higher than the adjusted martensite finish temperature, a full retransformation is not possible. This critical ambient temperature will not start a

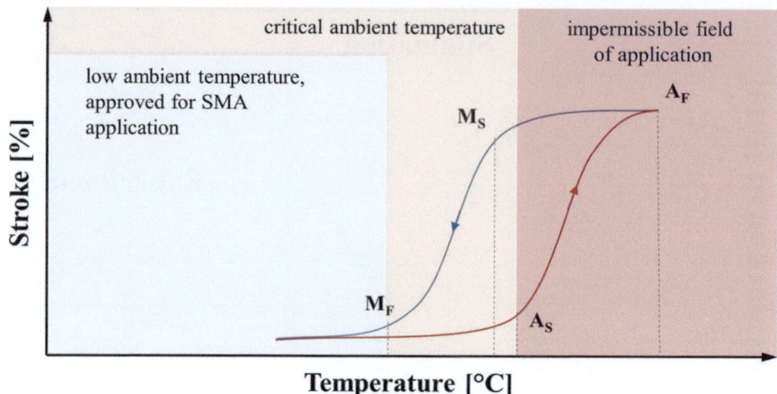

Fig. 9.7 Possible operating temperature field for electrical SMA actuators

transformation of the detwinned martensite into austenite, but if the critical ambient temperature will rise above martensite start temperature for example, the retransformation will not even begin. Nevertheless, operations in critical ambient temperature fields make still sense regarding one-way systems like safety valves for example which only have to fulfill a one-way function (for example to shut off a gas supply). If the ambient temperature is higher than the austenite start temperature, a thermal-induced transformation from martensite to austenite will be started regardless of electrical signal. At such high ambient temperatures, the electrical SMA actuator is not reliable anymore.

In order to estimate the operating temperatures, it is necessary to check the phase transformation temperatures of the used shape memory alloy material in combination with the calculated stress level. In practical experiments, thermal baths and climate chamber-based test rigs are used to validate the phase transformation temperatures and the system dynamics.

9.6.2 Electrical Activation

There are different approaches how SMA can be activated by the use of electrical energy. The most common method is the commonly described direct heating by Joule's law. Due to the variation of the resistance of the SMA material, at constant current level, the voltage level decreases with increasing volume ratio of the austenite fraction in the material. If you decide to set the voltage level to a constant value, the current will automatically increase. In practice a constant current level should be utilized in electrical support systems, because a current level change demands often more complex electronic system design in order to compensate energy fluctuations.

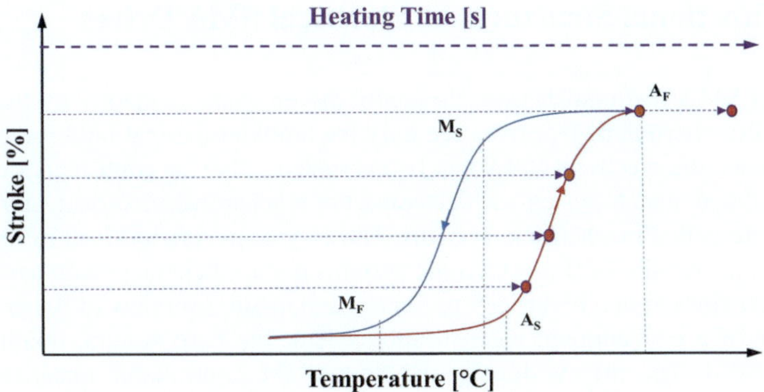

Fig. 9.8 The heating time in connection to the thermal SMA hysteresis

Besides the constant power input level, the activation time is vital for an SMA valve drive system's performance and the dynamic system response. Figure 9.8 shows the connection of the transformation curve to the heating time at a constant current level. The purple lines' length shows the heating time in order to reach a certain peak stroke. If the heating time rises, the temperature of the SMA element rises even over austenite finish temperature. This critical incident is called overheating and has several impacts on the SMA performance. First, it is unnecessary energy input and decreases the energy efficiency of the SMA drive. As a second it increases the necessary cooling time to reach the M_F temperature by the time necessary to cool down from the overheated temperature to A_F. Finally, as discussed in Chap. 3, it decreases the lifetime performance of the actuator by critical overheating resulting in damages on the structural level.

Of course, the heating time itself is also connected to the ambient temperature level. At higher ambient temperature level, the necessary energy input decreases due to the fact that the thermal level between the ambient temperature and A_S temperature is reduced. To summarize: if the ambient temperature rise it is necessary either to adjust the power input level or the power input time in the SMA actuator in order to achieve optimal activation. An enhanced insight in different control strategies is given in Chap. 5.

The optimal heating time can be obtained from experimental measurements, whereas variable factors like alloy type, heat-treatment type, element type, electrical activation type, stress level, and strain level are varied. For estimation of the A_F temperature either a precise displacement sensor is used (for detection of the SMA actuator's stroke) or the measurement of the intrinsic resistance values.

An alternative way for obtaining the optimal heating time in combination with the mechanical factors is by numerical calculation of the system which is described in Sect 9.5.

9.7 Functional Structures of Electrical SMA Drives

Electrical SMA distinguish from thermally driven systems mainly by the control logic: While thermal SMA drives use only the ambient thermal field for switching applications, the electrical SMA can be energized either by controllers depending on the actual demand. Systems which consist of mechanical, electrical, and IT components are called mechatronic systems. These systems, often driven by solenoids, are called in conventional mechatronic systems due to their successful assertion on the market since years. Figure 9.9 presents a schematic overview of the processing functions of a conventional mechatronic system: the base system, dominated by mechanical design, affects directly the flow of the controllable material. Within valve technology, this material is the fluid. A valve seat with the movable piston corresponds to the base system in Fig. 9.9. The actuator, classically seen as a part of the electronics is used to convert inserted power (e.g. electrical energy) into mechanical energy to change the base system's state. With this, also information (by electrical or mechanical means) is implemented into the state change of the base system. Another major part on the electronics level is the sensor element. This component detects the state of the base system by energetical or mechanical signals and converts it commonly into electrical information signals. As often described, the sensors are dominantly affected by the external surrounding in conventional mechatronic systems. The sensors have the function to generate electronic information which is leaded to a local signal processing unit. Either this information technology component is used to interpret the signal as a logical information or just to gain the signal and code it for submitting it via data-bus system to a main signal processing unit. The human interface is either connected directly to the local signal processing unit, or via a data-bus system.

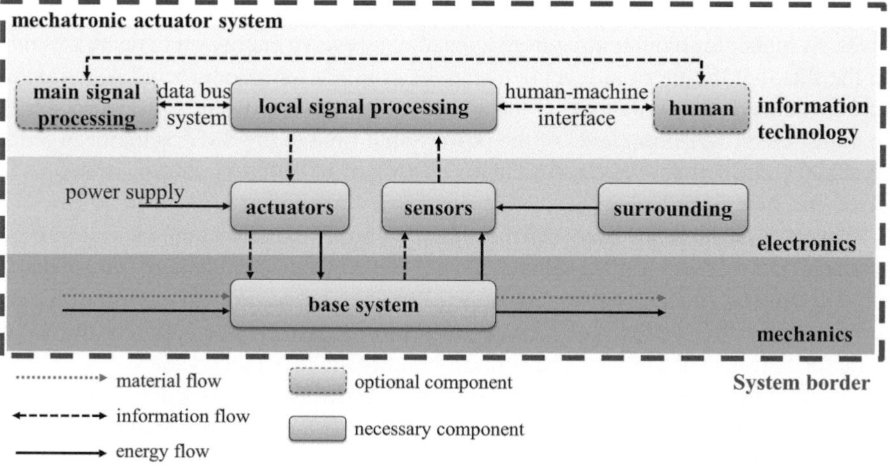

Fig. 9.9 Functional structure of a conventional mechatronic system in valve applications [7]

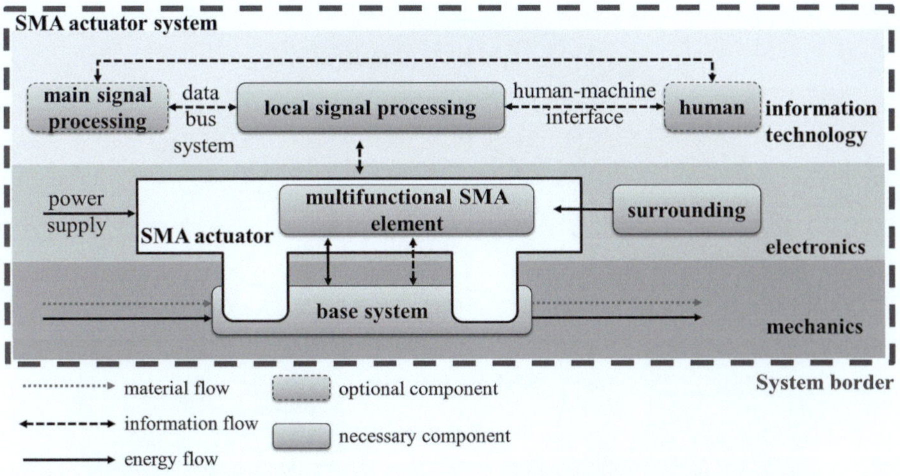

Fig. 9.10 Functional structure of a SMA system in valve applications

The main difference between the conventional mechatronic and SMA system focuses the SMA actuator itself. If even called SMA actuator, its core SMA element, is a multifunctional material. As discussed in Chaps. 3–5, in addition to the conversion of electrical into mechanical power, the SMA element has intrinsic sensor functions. This allows to use this element as actuator-sensor element substituting both separate elements of conventional mechatronic systems. Due to its mechanical design, even functions of the base systems can be assumed by SMA. Exemplary, a valve piston itself can be made out of SMA material (Fig. 9.10).

9.8 Component Structure of Electrical SMA Valve Systems

As often disregarded, the development of electrical SMA actuator systems does not only focus on the calculation of the SMA element's work output or dynamical design, but also on the system design. The SMA system consists, in the mechatronic system development of the integration of the SMA actuator into a device which will provide a function. The chart in Fig. 9.11 shows the periphery of an SMA valve system, consisting not only of the SMA element itself, but also on periphery needed to integrate the SMA element as well as used electronics and mechanics. Also valve-typical elements as valve seats and piston have to be regarded in the development of a SMA valve system. Other boundary solutions have to be found for constructive problems like power supply, control methods or activation strategies. In addition, it is important to regard the influences coming from the surrounding of the device which can affect the valve seat as well as the SMA drive element directly. The dominant influence from the surrounding of a SMA valve system is the thermal ambient field due to its direct impact on the SMA element's dynamic and mechanic characteristics.

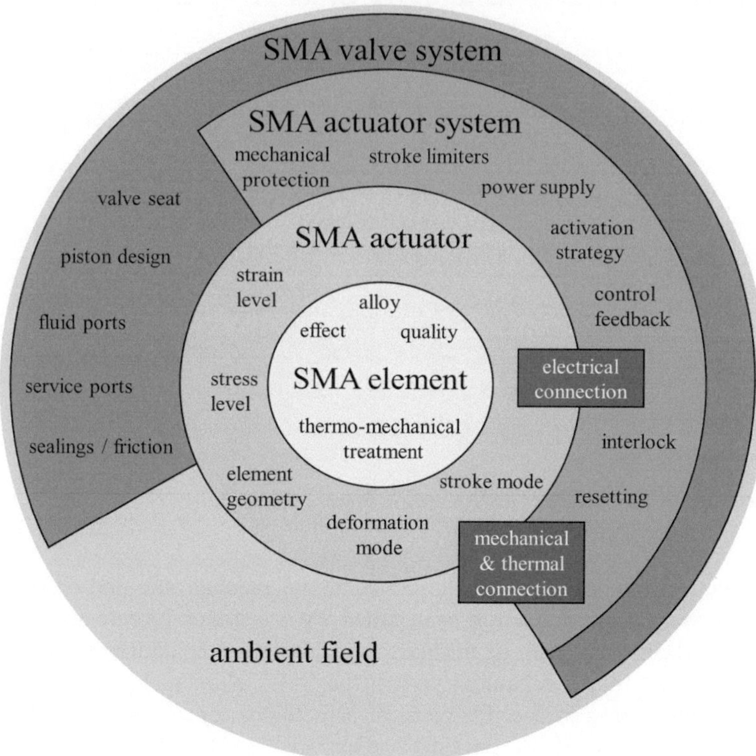

Fig. 9.11 Hierarchy of SMA valve system from the designer's perspective

Facing the development of an SMA valve systems a vast amount of parameters have to be discussed by the developing engineers. Due to the sensibility of SMA's characteristic on external parameters, the development cannot be accomplished step-by-step from the inner SMA element to the external systems because the system elements such as resetting or the valve seat (related to the fluid pressure) affect the SMA element. In future developments, a methodical approach will be used after guideline VDI 2248 will be published [8].

Generally, a development starts with the critical characteristics of the SMA actuator. The characteristics have to be checked with the SMA element itself, and the external hardware of the system. The goal of this level is to match if the needed characteristics can be fulfilled by SMA and to determine a general actuator concept. Therefore, the major requirements for actuation like the stress–strain levels and the element geometry have to be calculated iteratively either with the numerical simulation or the fast-track calculation. For the calculation of advanced systems, finite element analysis can be used for the calculation of complex 2- or 3-D SMA structures. A detailed and exact explanation of SMA modeling is given in [9].

Databases or the numerical solutions give the construction form, the alloy type, the actuation type, and the activation characteristics as output factors. Alternatively, these factors can be obtained from experiments. On the innermost level of the system, the SMA element's parameters like thermo-mechanical treatment, the material quality, and the used alloy (and effect) define the significant systems' parameters. Normally, these factors are set by the SMA element manufacturers.

Functional features like resetting, activation power, electrical supply, or mechanical protections against overstrain are summed up on the SMA actuator level. A selection in a wide range of concepts is possible on this level. For reasons of clarity, morphological tables can be used to estimate different solutions. Exemplary, it is possible to select from electrical activation (using the resistance-based self-heating) or by thermal fields or even by thermal generators (e.g. inductive heating). Mechanical features also can set an actuator's characteristics critically. Locking systems, as well as antagonist principles (two SMA working against each other) can be designed in systems which should hold certain positions and forces without the need of permanent energy consumption. The movement mode is also of major importance to the field of functional options. After proving of the maximal possible displacement, a stepped (several specified positions), controlled (linear movement with position controller), or binary actuation (only switching between two stable points) can be arranged. These functions can be achieved by the usage of fitted electronics. The time variant steering is the simplest method for activating SMA actuators for a certain time period. This sequential activation can be triggered by external signals (e.g. switches, end-stoppers). Integrating sensors it is possible to build position- and force-controlled systems. The hysteresis characteristic can be implemented in adaptive controller-electronics, which can be calibrated.

Developing the electronics should be done in co-work with software engineering. The fitted software is not only used to enable the control functions but can also be used for monitoring the system. From a simple counter which allows enumerating the used cycles up to systems which can perform a self-analysis procedures, software can provide these new functions. According to aspect of online communication, systems which report a malfunction (e.g. fatigued SMA wire) are also imaginable.

Other constructive options address in detail the sub-elements of shape memory actuators. The mechanical and electrical connection to the system influences not only the installation space and mechanical characteristics, but also the system's dynamic responses and the system's fatigue behavior. As shown in Chap. 4, the thermal influences of crimps on shape memory actuators can cause an imbalanced thermal field over the SMA element. If the system is heated, it will either work partly, or will be overheated in the midsection if the crimps are too grand in comparison to the SMA-element. Therefore, a crimp-specification list can be used to select empirically tested standard crimps and match them with the actuator's needs. Other connection methods like laser welding and polymer connecting can also be regarded in this phase. Fatigue prevention is also a major aspect on this development level. Stroke limiters for example can double the lifetime performance because additional elongation and structural failure is prevented. Other fatigue prevention elements, like re-stress elements are also imaginable in this phase. The consideration of integration aspects is finalizing this level. All functional and constructive options have to be checked for integration efforts. Exemplary, a locking mechanism has to fit in the installation space as well as the mechanical connection has to do. Both aspects have to fulfill criteria of cost, and automation possibilities in production processes.

The highest level in the SMA valve system model is the integration of the system into the imagined valve system. Here different aspects like mounting, sealing, integration material parameters, and boundary influences like vibrations and thermal fields are focused. Also production aspects like assembly, handling and testings should be considered. Another major aspect on this level is the balance between the systems' requirements and the preferences of the designed system. Here the critical needs have to be sorted into an assessment grid, which allows to score the achieved characteristics.

9.8.1 Stroke Limiters

Stroke limiters are used to specify the movement range of the SMA contraction during heating and elongation during cooling. By application of such limiters, additional elongations due to inhomogeneous cooling along the SMA wire structure are averted. If an assumed SMA wire cools down from the austenitic state at a constant load as presented in Fig. 9.12, the usable stroke decreases. Using a limiting element, for example a blocking element, no additional deformation of the SMA element over the adjusted strain can emerge (a–c). If such blocking element is not used, the cool parts of the SMA element at (c) will be overstretched gradually during cycling. Hence, additional elongation can be observed and misinterpreted as functional fatigue. Note that the degradation of the SMA element's stroke does not necessarily follow the graduation of the overstretching. During cycling, the overstretching can lead to constrictions of the SMA structure and to high peak stress levels in these points leading to mechanical fatigue of the SMA element.

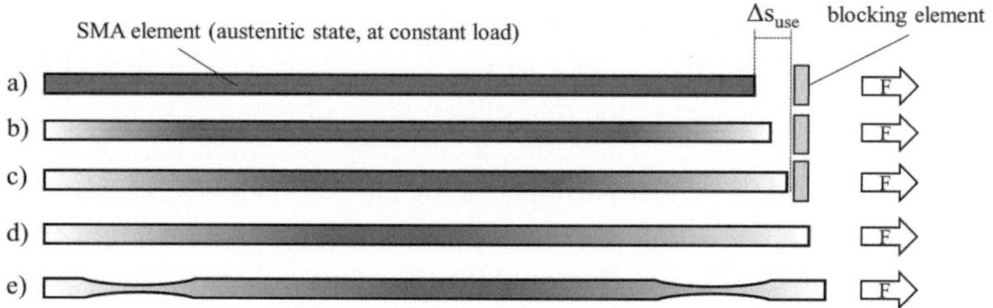

Fig. 9.12 Cooling of a pre-stressed SMA element ((**a–c**) with blocking element, (**d, e**) without blocking element)

9.8.2 Stress Protection

While stroke limiters protect the SMA element from overstretching, stress protectors limit the maximal stress levels which affect the SMA's cross section. The simplest method is to use a conventional spring element which starts the deformation at a force level which equals the maximal usable force level of the SMA element. If such additional or unexpected force occurs during cycling (for example by misoperation of the device), the stress protector element (for example tensile spring) is deformed after the maximal estimated stress level for SMA has been reached. Another example of a stress protector is presented in Chap. 10, Fig. 10.10.

9.8.3 Connection of SMA Wires

Among several concepts for the connection SMA wires to the mechanical and electrical systems, four preferred solutions can be distinguished according to the joining strategy (Fig. 9.13). On one hand, a force-fit joining can be used to compress the SMA wire in a mechanical element; on the other hand, material bonded joining methods can be used to integrate the SMA element in polymer structures or to weld them to other components. The connection technology does not only have a significant influence on the mechanical functionality, but also on the thermal properties of the system. Caused by a connecting technique the heat dissipation of the SMA element in turn affects the fatigue behavior of SMA actuators.

To ensure force transmission between the SMA actuator and the system environment, a suitable connection technique must be selected. Due to the already mentioned high relevance of connectivity for the entire system functions, the selection of the type of connection between SMA and the system plays a crucial role. Especially in electrically activated SMA valve systems where puller wires can be used, the connection between wire and a piston must be able to transmit high forces.

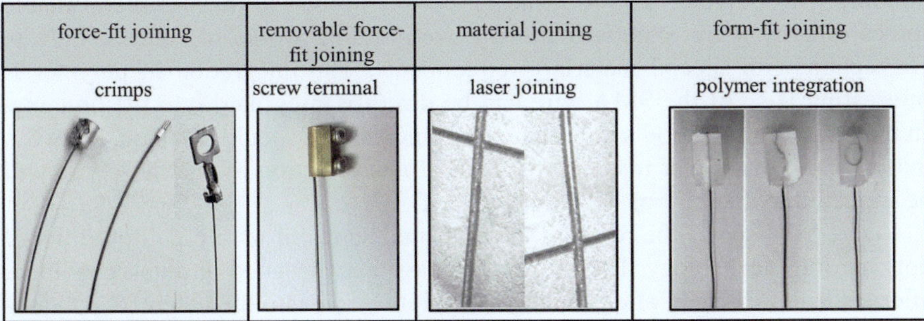

Fig. 9.13 Different connecting strategies for SMA wire actuators

The connection still is subject to significant fluctuations in temperature. Further, the transformation of the SMA element, which takes place within the connection system, has an influence on the durability, which has not yet been sufficiently investigated to date.

The connection by clamping or squeezing of the SMA wire to the mounting of the system also provides a very good connection. There a variety of material combinations are possible in this process, without affecting the durability. Ferrules made of copper or aluminum have in experiments, such as the following chapter shows, shown a very high stress potential so that this compound is very suitable for the connection of the SMA elements to peripheral systems. Also a strategy for mass production is given in this method. This is a reason why this method today plays a central role in the attachment of tension wires. However, it is important to ensure that the structure is not damaged by the joining process or altered in their properties.

For initial tests often compounds based on luster terminals are used. These are ideal because of their ability to disassembly, for simple experiments. Unfortunately, this method does not match the needs for serial applications due to poor reliability, reproducibility, and the possibility for automatic mounting solution in mass production applications.

The most important mechanical material connections is the welded joint. It is a permanent connection. This form of joining shows in connection with the same materials, i.e. in joining NiTi, with NiTi, the best results in terms of reliability. This procedure may be found, for example, application in the production of SMA-based shrink rings. However, also connection from heterologous materials is possible. NiTi alloys can be welded, for example with copper. A welded joint between NiTi and steels is possible. The problem, however, is the temperature input in the SMA element in welded joints. As a result, the functional properties can be significantly changed. With respect to the welding process, laser welding is particularly suitable for use in the SMA technology. One reason lies in the fact that the temperature entry through the laser affects only a very small punctual area.

Another possibility is soldering. As studies have shown, however, the mechanical strength of these compounds is low, which is why they so far do not apply in the industry. Also, the bonding is an interesting connection option.

Another possibility of joining, under the principle of form fit, provides overmolding with polymer. The form-fitting here is needed to transfer the available forces reliably. Pure material coherent overmolding works in contrast only in connection with special material pretreatments. The integration in polymer is interesting if so of the SMA wire can be directly integrated into the housing. In the future, this method's importance will increase. It should be noted, however, that during the polymer injection the phase transformation of the SMA wire will be triggered by the heat of the polymer itself.

Figure 9.14 shows the result of experimental series of tensile pull-out tests of polymer joint technology. The SMA wires have been inserted in initial condition, after a silane treatment, after sanding as straight wires into the polymers polypropylene (PP), polyoxymethylene (POM) and polycaprolactam (PA6). Additionally, the advantages in increase of max. tensile stress can be obtained from the combination

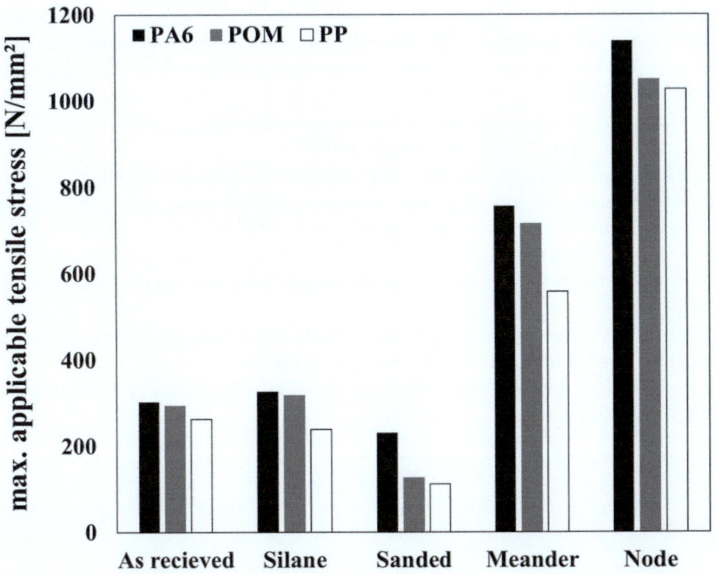

Fig. 9.14 Max. tensile stress levels for polymer-based connectors for SMA actuators

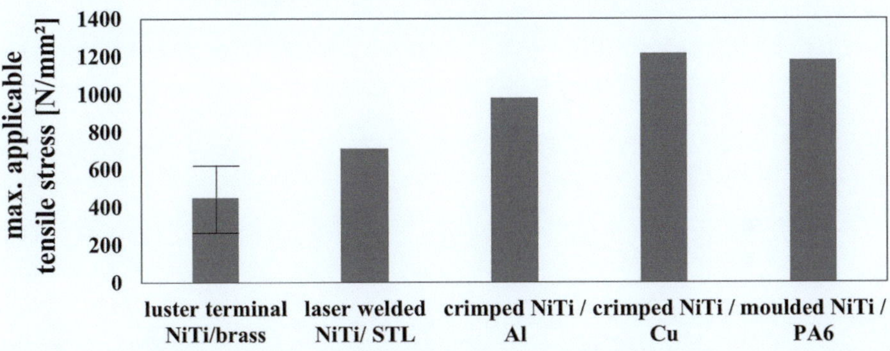

Fig. 9.15 Comparison of tensile stress levels of different SMA connectors

of material and form-fit as presented by SMA wires which have been inserted in meander and node form into the polymer moulds. These combinations enable the highest possible tensile stress levels (Fig. 9.15).

References

1. S. Langbein, A. Czechowicz, *Konstruktionspraxis Formgedächtnistechnik* (Springer Vieweg, Mannheim, 2013). ISBN 3834819573
2. A. Czechowicz, S. Langbein, J. Pollmann, in *Benefits of Standardization Illustrated by Shape Memory Actuators in Machining Applications*, VDE Kleinantriebe Tagung (VDE Conference on small drives) 2013, Nürnberg, Germany, September 2013

3. L.Oelschläger, *Numerische Modelierung des Akrivierungsverhaltens von Formgedächtnisaktoren am Beispiel eines Schrittantriebes (Numerical modelling of shape memory alloy actuators by example of a stepper motor)*, Ph.D Thesis, Ruhr-University Bochum, German, Shaker, 2004, ISBN 978-3-8322-2671-8
4. C.Liang, C.A. Rogers, One-dimensional thermomechanical constitutive relations for shape memory materials. J. Intel. Mat. Syst. Struct. **1** (1990)
5. F. Schiedeck, *Entwicklung eines Models für Formgedächtnisaktoren im geregelten dynamischen Betrieb (Development of a model for shape memory actuators at controlled dynamic movement)*, Ph.D. Thesis, Leibniz University Hannover, German, 2009
6. H. Meier, A. Czechowicz, in *Computer-Aided Development and Simulation Tools for Shape-Memory Actuators*. Metallurgical and Materials Transactions A, vol 42 (Springer, 2011), ISSN: 1073-5623
7. VDI Guideline 2206, *Development Method of Mechatronic Systems* (Beuth, Düsseldorf, 2004)
8. VDI Guideline 2248, *Shape memory technology* (in press)
9. D.C. Lagoudas (ed.), *Shape Memory Alloys. Modeling and Engineering Applications* (Springer, New York, NY, 2008)

Chapter 10
Methodology for SMA Valve Development Illustrated by the Development of a SMA Pinch Valve

Sven Langbein and Alexander Czechowicz

10.1 Motivation for a SMA Development Methodology

Because of the complex functional and hierarchical structure of shape memory alloys (SMA) and the vast number of different SMA-based valve drive solutions, many constructive ideas in relation to certain problems can remain hidden among different approaches during engineering. Owing to the complexity of development and an unfamilarity with this technology, especially in small and midsize companies, the cost of development is still very high. In Fig. 10.1, SMA development costs are compared with those associated with a solenoid device. Both devices have comparable characteristics (electrical valve application, actuator force of around 10 N, and a stroke of around 4 mm) [1]. Special emphasis can be placed on the first phase, the draft concept. In comparison to the well-known solenoid technology, the development of SMA systems, especially without a guiding procedure, can be expensive because of the longer development time. In this context, Fig. 10.2 shows some attributes of SMA valves that need to be considered during the product development process. Without methodological support or a deep knowledge of the process and technology, the development process can be completed only with great difficulty.

It is obvious that a system development procedure that uses solenoids is less expensive than SMA development. Many guidelines, norms, and standardized systems (actuators, models, and construction tools) are in place for the existing systems. With regard to SMA development, only a handful of numerical simulations exist

S. Langbein (✉)
FG-INNOVATION GmbH, Universitätsstr. 142, Bochum 44799, Germany
e-mail: langbein@gmx.com

A. Czechowicz
Forschungsgemeinschaft Werkzeuge und Werkstoffe e.V.,
Papenberger Straße 49, Remscheid 42859, Germany

© Springer International Publishing Switzerland 2015
A. Czechowicz, S. Langbein (eds.), *Shape Memory Alloy Valves*,
DOI 10.1007/978-3-319-19081-5_10

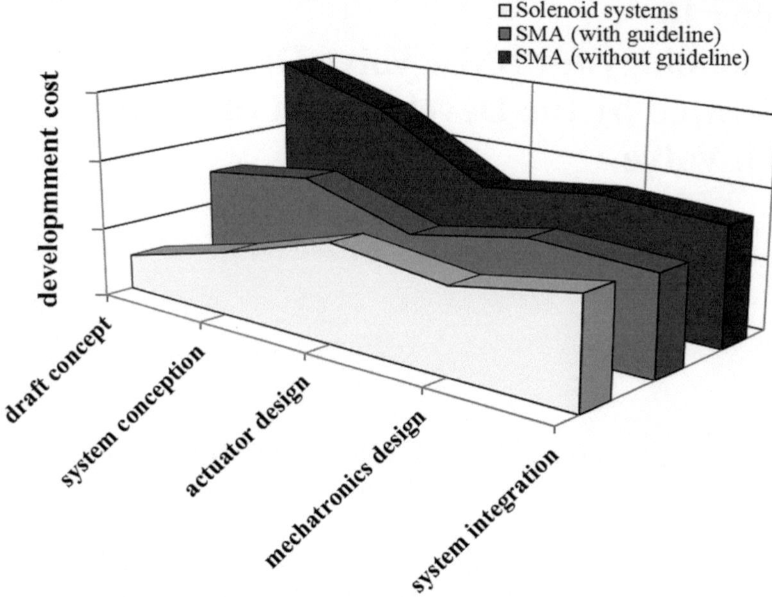

Fig. 10.1 Costs from SMA development process in comparison to solenoid actuators

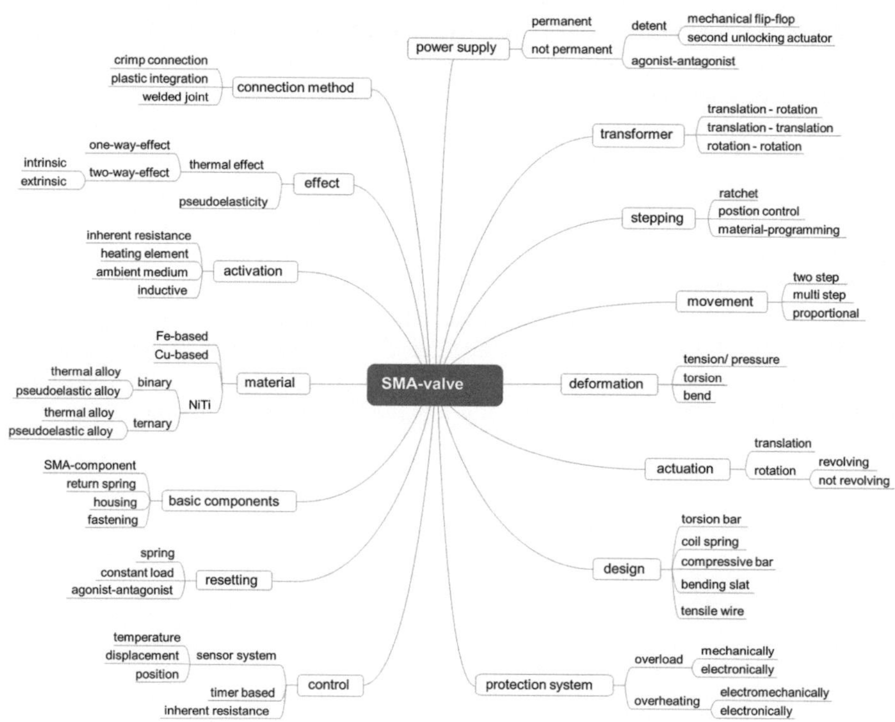

Fig. 10.2 Attributes of SMA-based valve systems

that can be easily used for concept calculation and optimization. Instead, a wide range of finite-element method (FEM) analysis tools for SMA are available that focus mainly on semifinished SMA components, but not on the system or the mechatronic elements. As a result of these factors, the development costs associated with shape memory actuators are very high at the beginning of the process. To enhance construction knowledge and reduce these costs, a methodical guideline leading the user through the possibilities of SMA step by step can be utilized for product development, as presented in the following section on a SMA valve. Figure 10.2 provides an overview of the attributes of a SMA actuation system. These attributes must be considered for the development of SMA valves.

10.2 Pinch Valve as an Example of a Methodical Development Process

The demonstrated development of a SMA-based fluidic valve illustrates the challenges associated with developing shape memory actuators. This simplistic design process may serve as an introduction to a constructive mindset in dealing with SMA.

To clarify the development of a simple SMA system, the development of a simple pinch valve is presented in this chapter. Among the different strategies used to control fluid flow, a possible solution is to squeeze a conduit tube. Such valves are often used in medical applications, where frequent changes of wetted parts are necessary, and for viscous fluids. Figure 10.3 shows an example of a solenoid-driven pinch valve currently available on the market [2].

For the utilization of SMA in a pinch valve application, a methodical approach to their development is needed. First, the requirements for the pinch must be clarified. The requirements for a pinch valve, apart from the compact, elongated shape, are a lighter weight in comparison with electromagnetic systems. Focusing on the mechanical part will allow us to provide an overview of the geometry, the mechanical mechanism of action, and the use characteristics of the valve. The overview of the mechanical requirements found in Table 10.1 includes the first rough specifications of the product to be developed. The information about the maximum operating pressure given here has a significant effect on sizing, especially in directly controlled valves. Thus, the analysis of the mechanical requirements can roughly address issues surrounding product dimensions and modes, usable materials, activation power/energy type used, and operating conditions. In Fig. 10.4 experimental data on a squeezed silicone fluid tube under different pressures show the required squeezing force range of the proposed SMA valve. At an estimated pressure of 3 bar, the actuator should exceed a closing force of 40 N.

The requirements for this relatively simple application can be obtained largely from the requirement specifications. These requirements already give a first indication of the design of the SMA element or the actuator system. Because the operability of the valve must be ensured at temperatures between −20 and +40 °C, for example, only a shape memory wire with phase transformation temperatures

Fig. 10.3 Solenoid-driven
pinch valve [2]

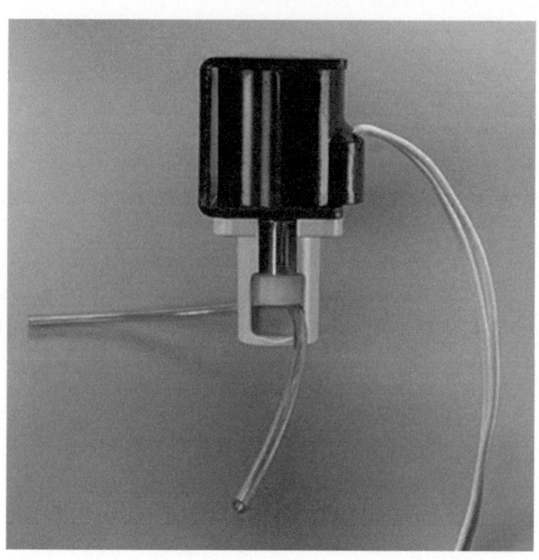

Table 10.1 Relevant requirements

Requirement	Value
Dimensions	Length <145 mm, diameter <30 mm
Drive	Shape memory alloy wire
Type	Direct, squeezing fluid guiding tube
Opening characteristic	Binary/optionally proportional
Valve operation	Normally closed, fail safe
Lifetime	>40,000 cycles
Maximum electrical energy	12 V, 1.5 A
Required stroke	4.1 mm
Spring constant	2.44 N/mm
Operating pressure	<3 bar (nominal 2.5 bar)
Dynamic properties	Activation <1.5 s, closing <6 s
Ambient temperature	−20 to +40 °C
Materials	NiTi, ABS, steel

over 40 °C can be used. If the phase transformation temperatures are below 40 °C, an incomplete retransformation of the actuator would occur during the resetting step. Additional boundary conditions for the actuator are the displacement, dynamics, lifetime, and mechanical load, which are connected through fatigue behavior. In this example, an output load of approx. 30 N must be generated to compress a pressure spring that normally quenches the tube. This value was obtained from the measurement curve in Fig. 10.4 at a pressure of 2.5 bar. The figure shows the force needed to quench the fluid tube under pressure.

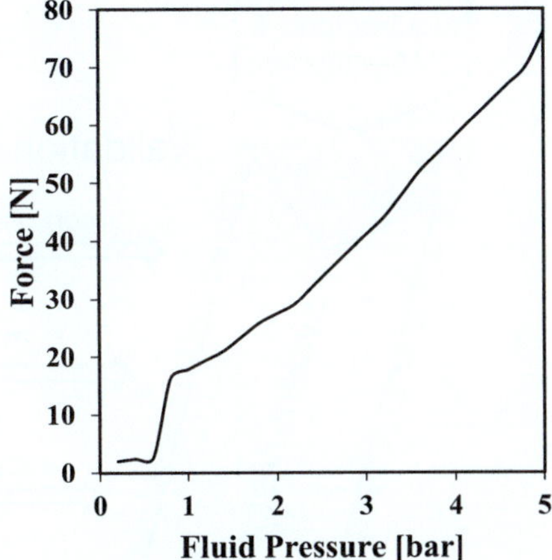

Fig. 10.4 Squeezing force required for closing silicone tube

The prestrain, which resembles a contraction, must be chosen in a range from 0.5 to 3.5 % iteratively by considering the stress levels and the expected lifetime performance of more than 40,000 cycles. Since no proportional pressure or volume control is required in the presented simple use case, only a digital switching mode is generated.

10.3 Methodology for SMA System Development

A methodical product development process offers a rapid and effective product development cycle through a structured procedure and through support provided by development tools. There exists a wealth of product development methods, such as the German engineering Guideline VDI 2221 for product development, which emphasizes mechanics, or VDI 2206, which focuses on mechatronic products. Since most of the many components of various domains, such as mechanical and electrical systems, are used in SMA-based products, an interdisciplinary methodology in the development of SMA valve systems is recommended. Generally, SMA actuators can be considered mechatronic systems that come with special requirements during the development process. Here mechatronics is understood as both a technical product type and an interdisciplinary engineering science that uses system techniques to integrate mechanics, electronics, and information technology and other areas of technology. The development of SMA-based products should be accomplished in accordance with those guidelines. But the use of SMA components in mechatronic systems is associated with new aspects not considered in previous process models. For this reason the development of SMA systems requires an extension or adaptation of the mechatronic process model. The so-called V-model for SMA actuator system development is shown in Fig. 10.5, which focuses on issues related to the development of SMA products. The presented V-model for SMA product development is part of the new VDI Guideline 2248 for the development of SMA actuators [3].

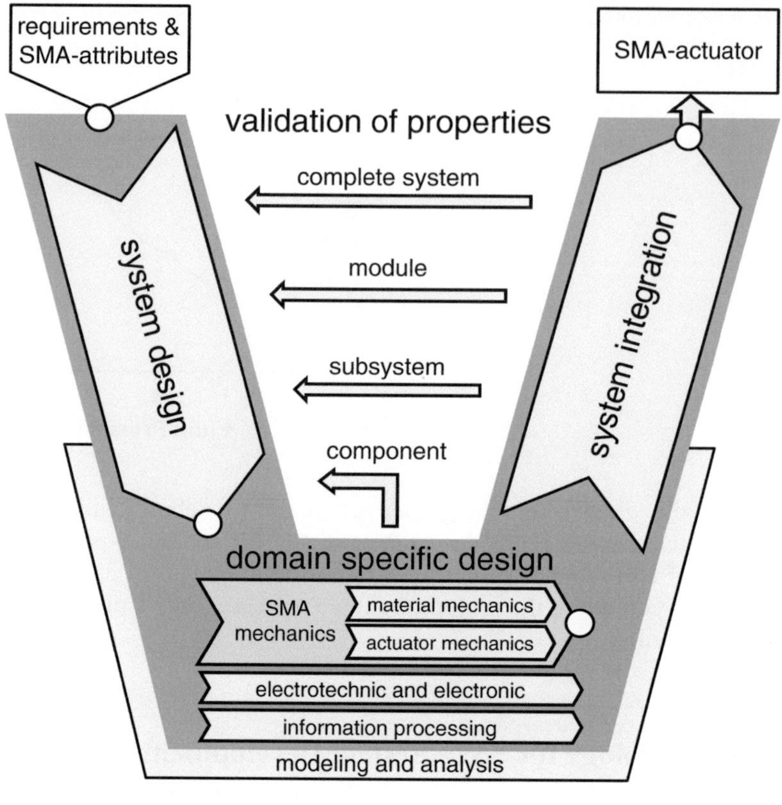

Fig. 10.5 SMA design methodology [3]

Idealized SMA development following the V-model begins with the identification of all requirements for the system to be developed. From these requirements the dominant parameters for SMA-specific functionality are elaborated in the second step. These parameters must be fulfilled by the SMA to achieve its technical potential. Following successful preliminary evaluation of the system's feasibility, a conceptual phase begins in the first system design phase. In subsequent phases, synthesis of possible solutions for the target system is elaborated with the aid of databases, experiential knowledge, or novel tools for SMA product development. In subsequent design phases, domain-specific subsystems of the actuator are designed. For example, SMA material mechanics and SMA actuator mechanics play a crucial role. The material mechanics describes how the material properties of the SMA element are determined. Examples include the alloy, thermomechanical pretreatment, and the associated hysteresis or transformation temperatures. Also, training should be addressed in this step. It plays an important role in terms of the durability and stability of shape memory effect. In the domain of actuator mechanics, the integration capabilities of the SMA actuator are highlighted as a priority in the overall system. Here, connectivity and power transmission are given special emphasis. Also, electrical engineering and information processing aspects were added to the SMA V-model because they are important in the development of all electrically activated SMA-based actuators. However, they can be ignored in temperature-sensitive and autonomous applications such as thermal valves. Following the successful design of domain-specific aspects, the developed subsystems are validated in experiments and then, during system integration, implemented in the overall SMA valve system. A quality gate closes in every phase. The quality gates may have different criteria for evaluating the requirements in relation to the attainable SMA properties. By using such a method, the iteration of the development process and thus the development time can be significantly reduced. Sometimes the requirements are incompatible with the SMA properties. In this case, system developers should abort the development.

10.4 System Design of SMA-Based Pinch Valve

System design is the most significant step in the development of a SMA valve system. In system design, initially the product requirements should be specified to clarify the basic functional relationships. On this basis, the search for the main function and various derived subfunctions is carried out, and they will be represented in a functional structure. The relationships between individual functions describe the energy, material, and signal flow within the system. The functional structure serves as the basis for the search for active principles that realize the functions at the physical level. In technical systems, the cause–effect correlation can also be supplemented directly by solution elements. For example, a distance measurement can be specified directly by the use of a sensor device if its use is advantageous for the application. Assigning active principles or solution elements to the solution-independent functions makes it possible to produce various solution alternatives. The system design is complete with the derivation of a solution-oriented building structure, which depicts all subsystems at the active principle level. The general procedure in the design of SMA-based systems is schematically shown in Fig. 10.6. The individual steps are further detailed and based on the following application example of a pinch valve.

10.4.1 Step 1: Preliminary Feasibility Assessment

The preevaluation aims to assess the feasibility of the project proposal in advance of the product development process. The review will be based on some key exclusion criteria of the project. In this way, project risk will be revealed and unnecessary effort spared. Using various criteria, the use of SMA technology is assessed on the

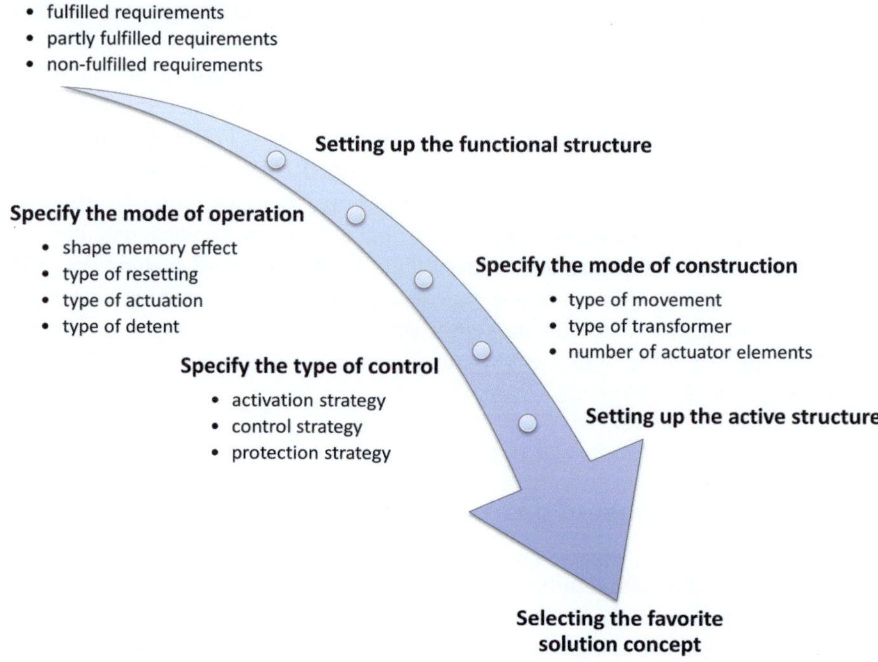

Fig. 10.6 System design procedure

basis of a comparison of technical requirements and attributes of SMAs. This step is the first quality gate. For this reason, this quality gate will serve as the foundation of the entire development project. One way to assess feasibility is through a systematic classification regarding the fulfillment of product requirements. The method for doing this is illustrated in Fig. 10.7. The requirements result from the product specifications and must be classified for each system. The requirements do not belong to the technology itself, but they influence the parameter values of the technology or the implementation of the technology. Generally, three main categories of requirements can be distinguished: fulfilled, partially fulfilled, and unfulfilled requirements [4]. In the following examples, marking the requirements with various smiley faces is a clear and simple way to classify product requirements.

10.4.1.1 Fulfilled Requirements

Fulfilled requirements can be divided into conventional and new requirements. Conventional requirements should not be viewed as a problem for the implementation of new technologies; at the same time, there are no decisive product advantages and innovations associated with these requirements. In contrast, the fulfillment

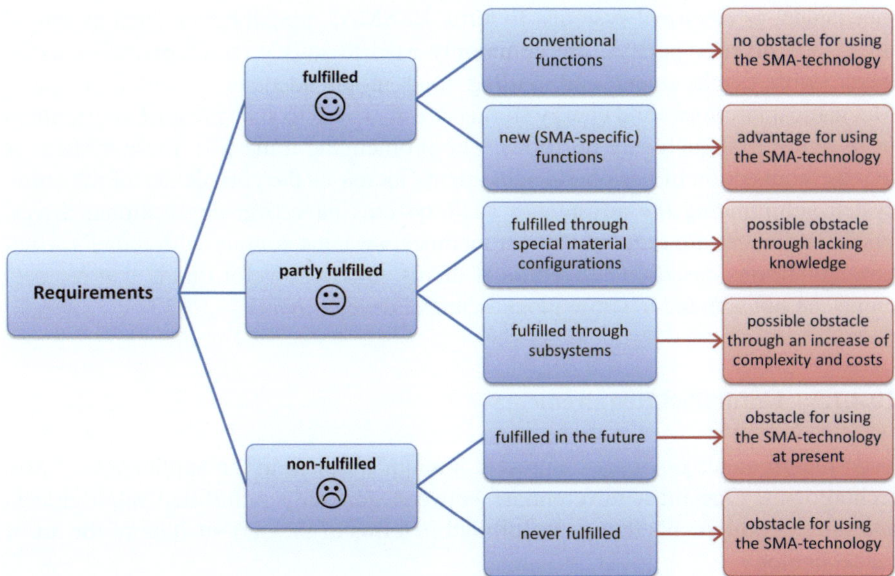

Fig. 10.7 Classification of requirements for SMA applications [4]

of new requirements should be seen as the main advantage of new technologies. These new technologies are what make product innovation possible at all. Shape memory technology involves the following new requirements: installation space, weight, sound emission, and system simplicity. Furthermore, new requirements can be demanded in connection with functional integration. In particular with the application of SMA, highly integrated components can be generated. Such components are detailed, for example, in [5].

10.4.1.2 Partially Fulfilled Requirements

Partially fulfilled requirements represent the first obstacle in the application of new technologies. Standard SMA materials cannot fulfill these requirements. There are two ways of satisfying such requirements. First, specific configurations of the material can provide the desired quality of a function. In the case of shape memory technology, special material properties, like the R-phase transformation, should be mentioned. The temperature hysteresis can be reduced from approx. 20 K to approx. 2 K by conversion. In this way, temperature controls can be accomplished more accurately. The thermostatic valve applications, explained in the preceding chapters, depend on this low hysteresis. Applications based on these special material configurations are hard to find owing to a general lack of knowledge. Second, unfulfilled requirements of a material can be fulfilled by connecting additional mechanical or electrical systems. In terms of SMAs, transformers, such as sets of levers and diverter pulleys, are commonly used to satisfy requirements in accordance with displacement and pulling load in an adequate installation space. Mechanical interlocks and energy storage are also used to keep positions currentless or to return to default safe positions. The problematic issue with these systems is that the connection of peripheral subsystems increases the complexity of the entire system, eliminating the advantages such systems have over conventional drives. Furthermore, certain requirements of technical systems comply with today's structural conditions based on conventional drives. One reason for this is that requirements are not adapted to subsequent technical developments.

10.4.1.3   Unfulfilled Requirements

Such requirements obviously represent a major obstacle in the application of new technologies. One must differentiate between generally unfulfilled requirements and those requirements that go unfulfilled in terms of the current state of the art of the technology.

With regard to the second group, a further technical development must take place. Sometimes, laboratory samples are available, but no industrial-scale implementation has been accomplished yet. In this case, however, product development should continue up to the first solution concepts. Depending on the solutions, filed patents should be checked. Moreover, the further development of the technology should be observed. Positive results should allow a resumption of development. In the case of SMAs, high-temperature alloys must be mentioned as part of the requirements to be fulfilled. Such alloys have transformation temperatures above 200 °C and therefore are mostly suitable for applications in vehicles. Owing to the poor formability of these alloys, semifinished products, such as wires, are unavailable. However, new alloying elements will solve this problem in the future. Nevertheless, such further technological developments require an enormous amount of research until a safe and reliable application can be guaranteed.

With regard to current and future unfulfillable requirements, the development process must be stopped; development should be continued with a different technology.

10.4.1.4   Requirement Classification for Pinch Valve

For the development of the pinch valve the SMA-specific requirements are classified in Table 10.2. The requirement of a closing time of 6 s given an ambient temperature of 40 °C have been identified as critical for a SMA valve drive. In consideration of the expected wire diameter, the cooling dynamic should be tested. In addition, the proportional opening and electrical requirements are critical because subsystems are needed. But no requirement is an obstacle to using SMA components in the valve system. For this reason the development process can continue.

Table 10.2 Classification of requirements for SMA-based valve drive

Requirement	Value	Classification
Dimensions	Length <145 mm, diameter <30 mm	☺/☺
Opening characteristic	Binary/optionally proportional	☺/☹
Valve operation	Normally closed, fail safe	☺
Lifetime	>40,000 cycles	☺
Maximum electrical energy	12 V, 1.5 A	☹
Required stroke	4.1 mm	☺
Spring constant	2.44 N/mm	☺
Dynamic properties	Activation <1.5 s/closing <6 s	☺/☹
Ambient temperature	−20 to +40 °C	☺

10.4.2 Step 2: Functional Structure

When the functional structure is prepared as shown in Fig. 10.8, the SMA-specific functions can then be incorporated into the system design. A particular point is the activation of the shape memory actuator. One of the valve's electrical circuits or a medium-driven opening can be used. Thus, the valve can open by itself if the fluid temperature gets too high. A shape memory can thus operate in this case thanks to its specific ability to adjust the electrical function and to serve a thermal emergency function. Two central roles of the functional structure are as a so-called fluid switch and an energy converter. The latter function refers to the SMA element and the electrical activation in a two-step process. First, the electrical energy should be converted into thermal energy and then into mechanical energy. With the safety function that governs high fluid temperatures, the thermal energy is directly converted into mechanical energy. The mechanical energy is then used to switch the fluid flow. Just as the actual valve function represents the system, resetting should be carried out by mechanically restoring elements automatically during deactivation/cooling. In the functional structure, the resetting element is seen as the storage of potential energy. This potential energy is released during cooling and reconverted back into mechanical energy, which turns off the fluid flow, and the valve closes. Simultaneously, the mechanical energy is used for stretching the wire again. The two processes of activation and deactivation, opening or closing the valve, are recorded in two different functional structures.

10.4.3 Step 3: Mode of Operation

The entire process of the phase in which the mode of operation is specified is shown in Fig. 10.9. To clarify the methodical concept clarification, the steps for specifying the active principles are presented in the following figures. The chosen pathway is

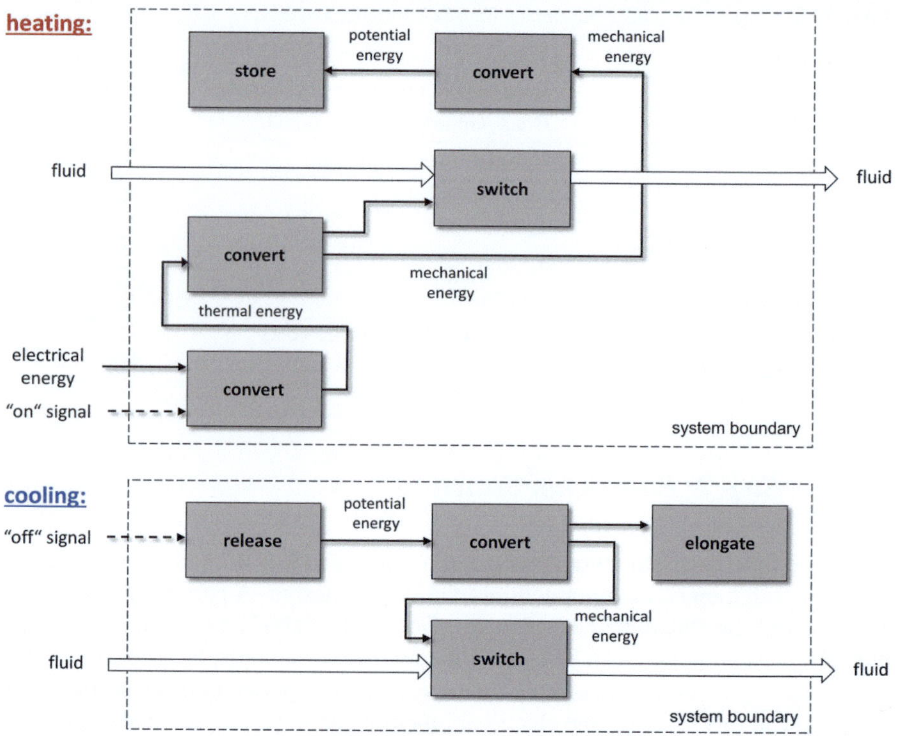

Fig. 10.8 Functional structure of valve system

always indicated by a check mark. Introducing this phase, the right effect selection leads to manually resettable (or single-use) SMA actuators, or usage of the extrinsic two-way effect. The requirements demand a fail-safe system, meaning that during the inactive phase of the SMA element, the valve is either closed or in the process of closing. To close the valve and reset the SMA, a resetting element is required. At this point it is possible to decide between an agonist–antagonist structure, where two SMA elements operate alternately against each other, a spring element, described earlier, and a constant resetting mass. As clarified in the requirements, a resetting spring is used for closing the valve. For detailed construction and spring-type selection, the domain-based design is discussed in the next chapter. A selection of restoring elements is presented in Table 10.3.

Focusing on the actuating behavior, it is possible to decide whether the valve should operate as a binary (open–close) valve, a stepped motion valve with a limited number of opening and closing steps, or as a proportional valve. Because a pinch valve should be used as a time-based dispersing valve, a binary behavior is selected. For further conceptual extensions, it is also possible to integrate a proportional controller based on the intrinsic sensor effect (Chap. 5), even if no sensor element is included.

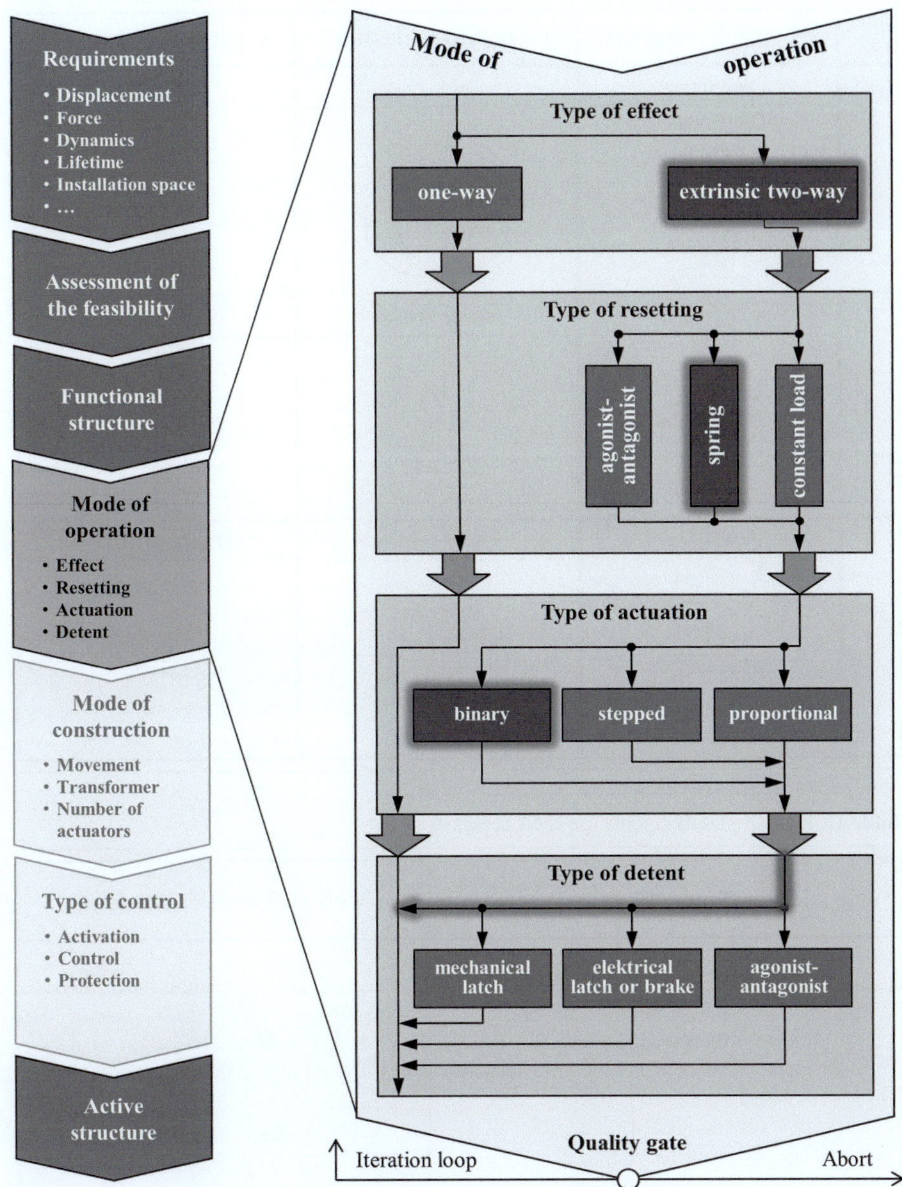

Fig. 10.9 Methodical system design – determining mode of operation

The last step in the functionality definition aims at the possible locking mechanisms. Table 10.4 presents commonly used locking mechanisms, which can be combined with SMA actuator systems. On the one hand, it is possible to lock or unlock
a mechanical lock by a second (unlocking) SMA actuator. Different constructive

Table 10.3 Overview of different resetting elements and resetting strategies for SMA actuators

	1 spring elements	**2** polymer elastics	**3** antagonistic SMA
1	coil spring 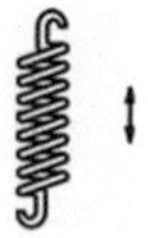	elastic rings 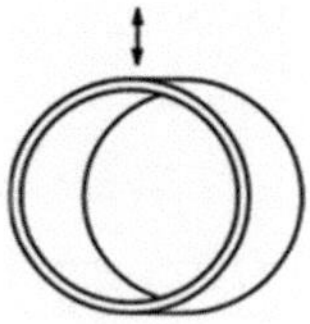	SMA wires 
2	spiral spring 	elastic ribbons 	SMA springs 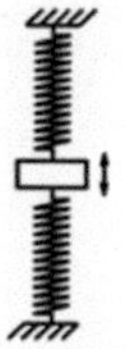
3	leaf spring 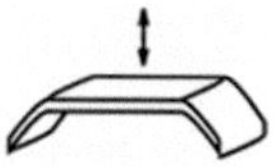	rubber bellows 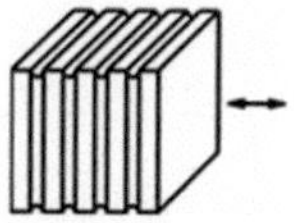	SMA sheets / ribbons 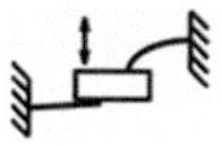

Table 10.4 Examples of detents for SMA actuators

with additional unlocking SMA actuator		with push-pull mechanism	
linear	rotary	linear	rotary
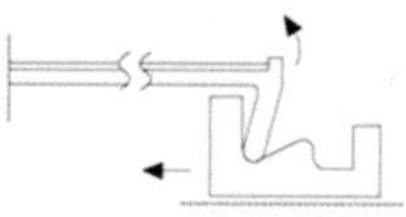	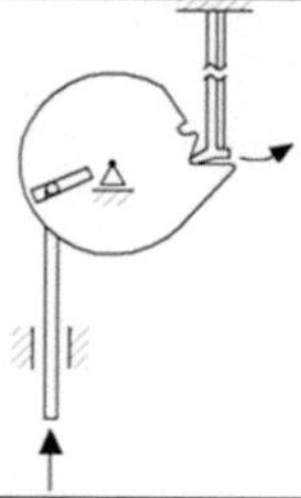	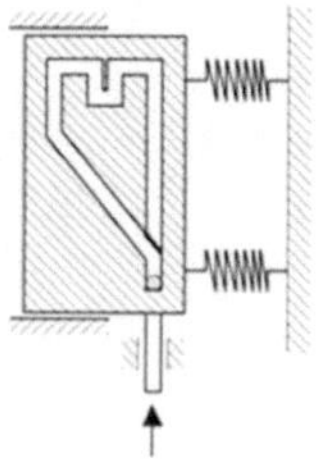	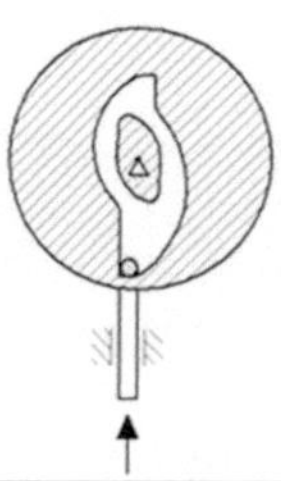
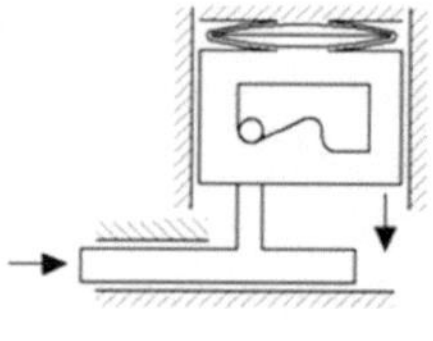	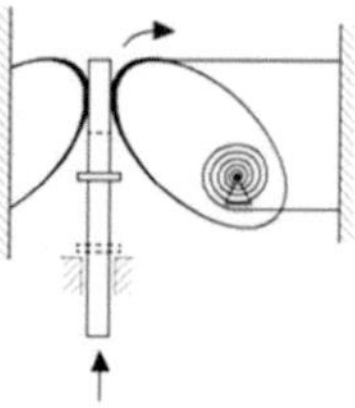	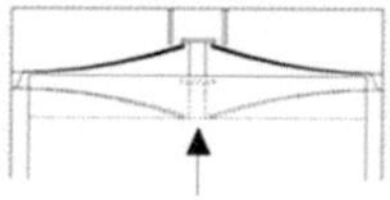	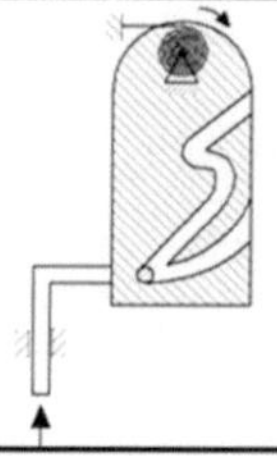

forms can be imagined, either a linear design or a rotary arrangement. On the other hand, a push–pull mechanism would be comparable to well-known elements from ballpoints. The major difference between the two principles lies in the cyclic dynamics and the design effort. Through the use of an additional unlocking actuator, the major SMA drive can be released after a complete transformation. Thus, the locked actuator output has no connection to the major SMA drive. If the minor SMA actuator is activated, the latch mechanism opens, and the system is reset. The return of the whole system is not determined by the cooling curve of the major SMA drive. In systems using push–pull mechanisms, the major SMA drive is activated for a second time to unlock the system. For this action, a cool-down time must pass.

The example development demands a fail-safe function, so that the system will not lock into any position and will return without energy input to the closed state. For this reason a detent is not necessary.

10.4.4 Step 4: Mode of Construction

In the next phase of system design, the SMA actuator design (focused on the actuating properties) is emphasized. Figure 10.10 shows the detailed design specification step of the SMA development method in which the mode of construction is specified.

At first it is necessary to choose the right movement type. One advantage of SMA actuators is that linear movements are easily generated using a linear wire, rod, pipe, sheet, and spring elements. These actuators can work under tensile load. Excluding the wire and sheets, these elements can also work under a compression load. A SMA rod, for example, will return to its original shape after reaching the austenite finish (A_F) temperature. Bending and torsional loads on SMA elements are also possible. To generate rotary or bending movements, spiral springs or sheets can be used. Table 10.5 provides an overview of different SMA elements in relation to the type of movement. In general, it can be asserted that in electrically activated systems, mostly wire actuators are used. Most coil springs are used in systems that are thermally activated by the surrounding medium. Coil springs have the advantage of compact dimensions and simple system integration without complex fastening technology. Wire actuators, in contrast, are very energy efficient and have better dynamics. If SMA actuators are also heated by their own resistance, only wire actuators provide an optimal solution because spring actuators require too high a current owing to their large cross sections.

For a pinch valve it is necessary to have relatively short activation and deactivation times. As discussed earlier, the cyclic dynamics of electrical SMA elements depends on the elements' surface-area-to-volume ratio. Because of the dynamics, the proper choice for the elements is a wire material. Because of the moderate forces resulting from the elasticity of the squeezing tube and the fluid pressure, a thin wire can be used.

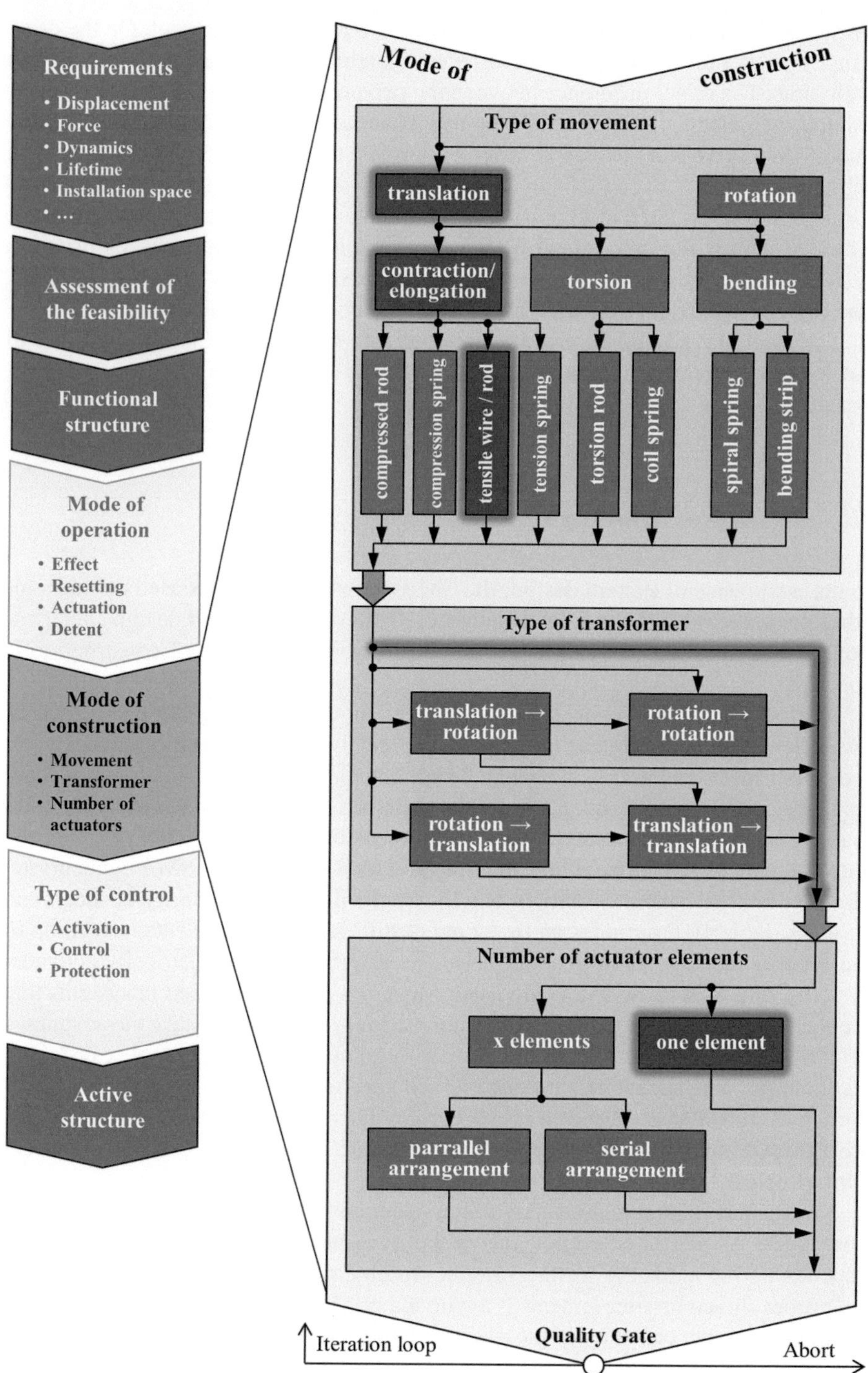

Fig. 10.10 Methodical system design – determining mode of construction

Table 10.5 Types of movement for different SMA elements

		Kind of Movement			
		Translation		Rotation	Bending
		Compression	Tension		
Form of Actuator	Spring	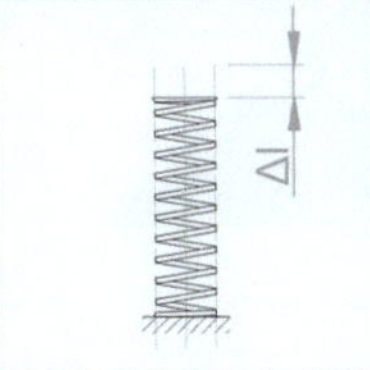	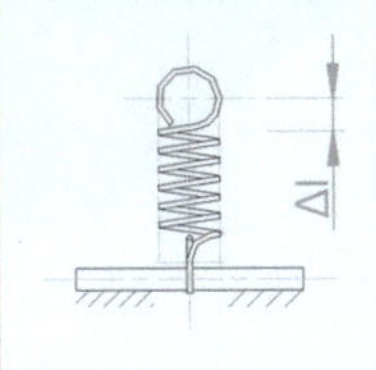	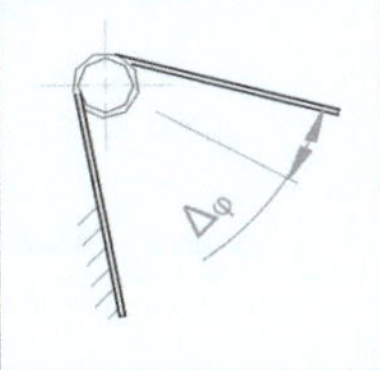	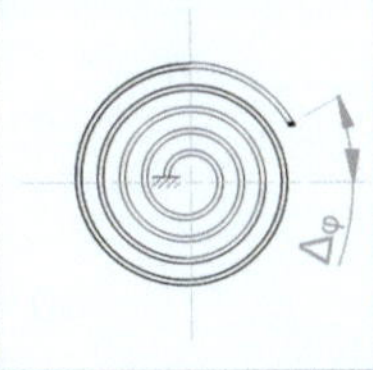
	Wire/ Beam	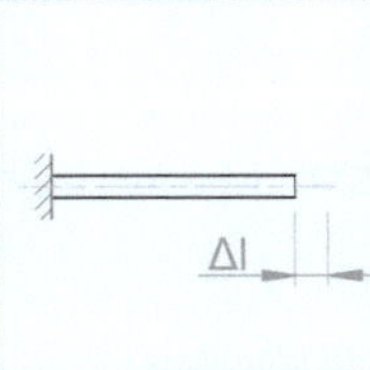	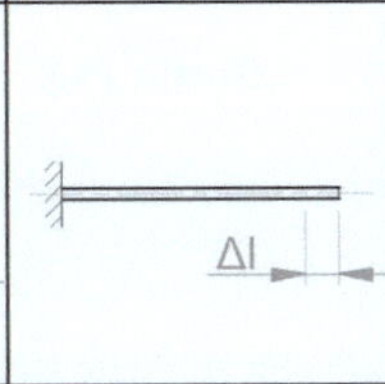	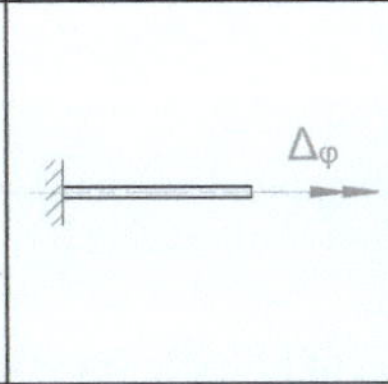	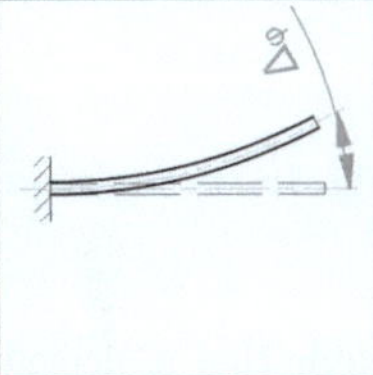
	Sheet	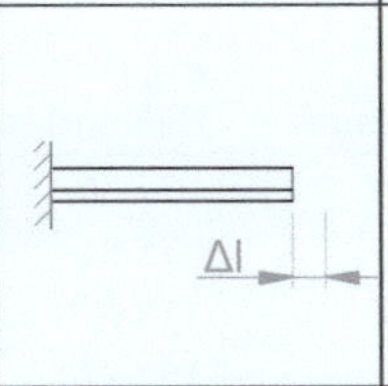	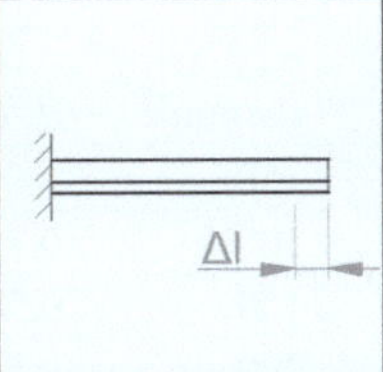	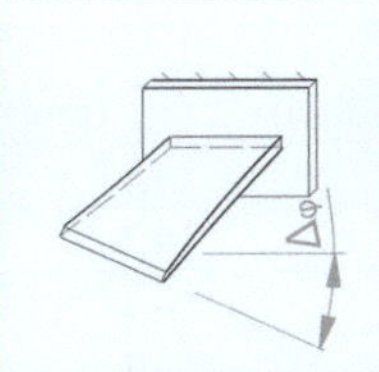	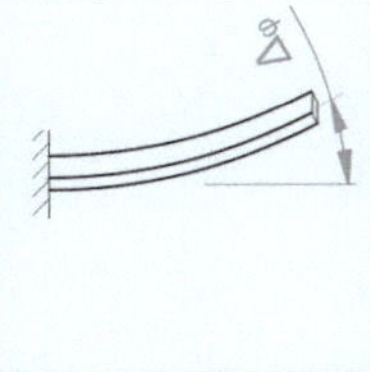

In this step, a fast-track calculation of the required diameter and the length is helpful. Since the lifetime is a relevant factor, this has to be included for the estimation of the maximum stress and strain in the calculation. Using the lifetime diagram presented in Chap. 3 (Fig. 10.11), it is possible to define stress–strain combinations for a system's design. The example presented uses a 3 % applied strain and a maximal load of 400 N/mm².

With the fast-track calculation (Chap. 7) it is possible to calculate the solution field for different stress and strain levels that are lower than the maximal values. With the spring constant $c = 1.7$ N/mm and the closed level force $F_{min} = 30$ N, the possible wire diameters (d_w) can be calculated for different applied maximal stress levels. These are presented in Table 10.6. At a maximal stress level of 400 N/mm² the SMA pinch valve requires a diameter of 0.36 mm and is prestressed at 294 N/mm². With respect to safety factors and common material availability, the diameter will be set to 0.4 mm. These values can be interpolated from Table 10.6.

Regarding the SMA wire length, the lower part of Table 10.6 shows a field of possible solutions. Depending on the required fatigue life, the maximum strain of 3 % has been determined by the diagram in (Fig. 10.11). By means of this strain the length of the wire can now be calculated. This results in a contracted wire length (length of the wire in the austenitic state) of 136.7 mm.

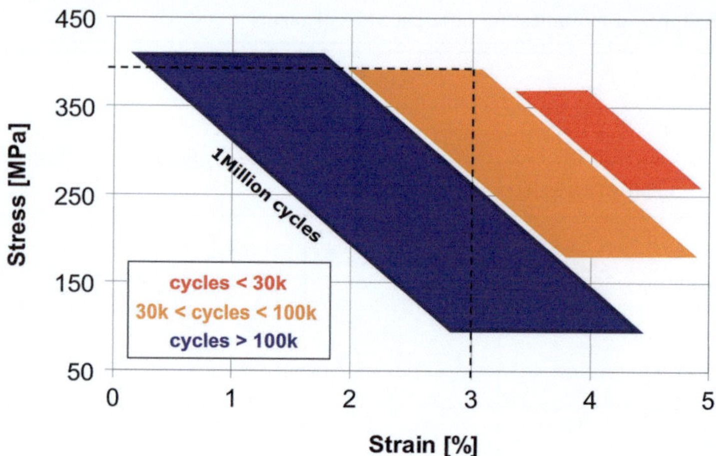

Fig. 10.11 Selection of applied maximal stress–strain levels on SMA wire for pinch valve

Table 10.6 Calculation of possible stress diameter and wire length solutions

F_{min} (N)	F_{max} (N)	c (N/mm)	s_{use} (mm)	σ_{max} (N/mm^2)	d_w (mm)	σ_{min} (N/mm^2)	σ_{use} (N/mm^2)
30	40	2.4	4.1	50	1.01	37.46	12.54
30	40	2.4	4.1	100	0.71	75.81	24.19
30	40	2.4	4.1	150	0.58	113.6	36.4
30	40	2.4	4.1	200	0.5	152.87	47.13
30	40	2.4	4.1	250	0.45	188.72	61.28
30	40	2.4	4.1	300	0.41	227.34	72.66
30	40	2.4	4.1	350	0.38	264.66	85.34
30	40	2.4	4.1	400	0.36	294.88	105.12

s_{use} (mm)	ε_0 (%)	ε (ε_0/100)	l_0 (mm)	l_1 (mm)
4.1	0.5	0.005	820	824.1
4.1	1	0.01	410	414.1
4.1	1.5	0.015	273.3	277.4
4.1	2	0.02	205	209.1
4.1	2.5	0.025	164	168.1
4.1	3	0.03	136.7	140.8
4.1	3.5	0.035	117.1	121.2

With these values it is possible to start the specification of transformers to optimize the force-stroke ratio of the designed SMA actuator.

With transformers as drawn in Table 10.7, the force or stroke level can be fitted to the specifications of the SMA actuator system. For force or stroke augmentation, a good choice is to use lever systems. Often the lever system is also realized in the form of a gearbox. According to the definition of lever arm length, the translation

Table 10.7 Examples of different transformers

1	force augmentation	2	stroke augmentation
	block and tackle system		pulley system
	SMA		SMA
	lever system		lever system
	SMA / F / 2F		SMA / l / 2l

Table 10.8 Examples of different augmentation arrangements for SMA actuator wires

1	force augmentation	2	stroke augmentation
	parallel arrangement		serial arrangement
	SMA		SMA

factor of the force and stroke can be adjusted. Among the different transformer types, especially in unlocking applications where the smallest stroke levels are used to release a high force brake.

Because of the calculation made in Table 10.6, no further transformers are needed for the pinch valve example, leading in the development method to a second option for augmenting either the force or stroke of the considered SMA actuator. In addition to the possibility of using transformers, augmentation of the force or stroke can be achieved by the utilization of several SMA elements. Table 10.8 shows the two possibilities for arranging SMA elements, represented by SMA wires. In a parallel arrangement the pulling forces of the SMA elements are added with a consistent stroke, while in serial arrangement the stroke increases at a consistent force.

In the example development, on the one hand, no force or stroke augmentation is required, as calculated in Table 10.6. On the other hand, regarding the link between the dynamics, the SMA wire diameter, and the force, it is necessary to reconsider the arrangement in the imagined concept.

10.4.5 Step 5: Type of Control

The fifth step of the system design concerns the control specifications. In this step (Fig. 10.12) the developer should consider the proper activation methods for the system. This step is executed after the fast-track calculations in the design specification step, in case no solution to the direct electrical heating issue can be found.

Direct electrical heating common method for SMA activation is not the only way to control the phase transformation. Table 10.9 presents a brief overview of the commonly used methods for SMA actuator activation. Alternatively, it is possible to induce thermal energy into a metal housing, an adjacent metal component or the SMA element directly by electromagnetic induction. Another possibility is the use of an external heating coil element. This solution is often used in a large cross section, as can be found in helical spring actuators. Often, an ambient medium heating medium is used in applications. The last option can be combined with electrical activation options, for example, as a safety function. For instance, a SMA element can work at −20 to 45 °C as an electrical closing valve. If the ambient temperature rises above these critical values, above the austenit-start (A_S) temperature, the valve could perform a safety closing operation triggered by the ambient medium. In the example, the pinch valve uses a resistance heating activation.

Following selection of the activation, a controller type should be selected. In Chap. 5, controllers with intrinsic sensor functions are discussed in a more detailed way. In brief, it is possible to steer an actuator by a time-variant ON/OFF switch using constant pulse times, a thermal controller measuring the temperature of the SMA element, a position controller device measuring the actual movement of the SMA element, and the intrinsic sensor function, which resembles linearly the phase transformation as presented in Chap. 5.

Because no disturbances and shifting thermal fields are expected, a simple time-variant controller can be selected for the example.

The last element to plan in the system design is an optional protection element. During the lifetime of a SMA system, several errors may occur. For example, additional friction may result from the corrosion of a piston or if a user makes a critical mechanical error during the application phase. Regarding SMA actuators, this could lead to an overload of the SMA element itself, which might affect the prestress of the system as well as its lifetime performance. To address these problems, mechanical protector elements can be incorporated into the system, as illustrated in Fig. 10.13. In this concept, a SMA wire works against a mass and the fixed bearing is embodied as a stiff tension spring. When the SMA is activated, it pulls the mass up over the usable displacement. If the SMA is prestressed by a significantly higher

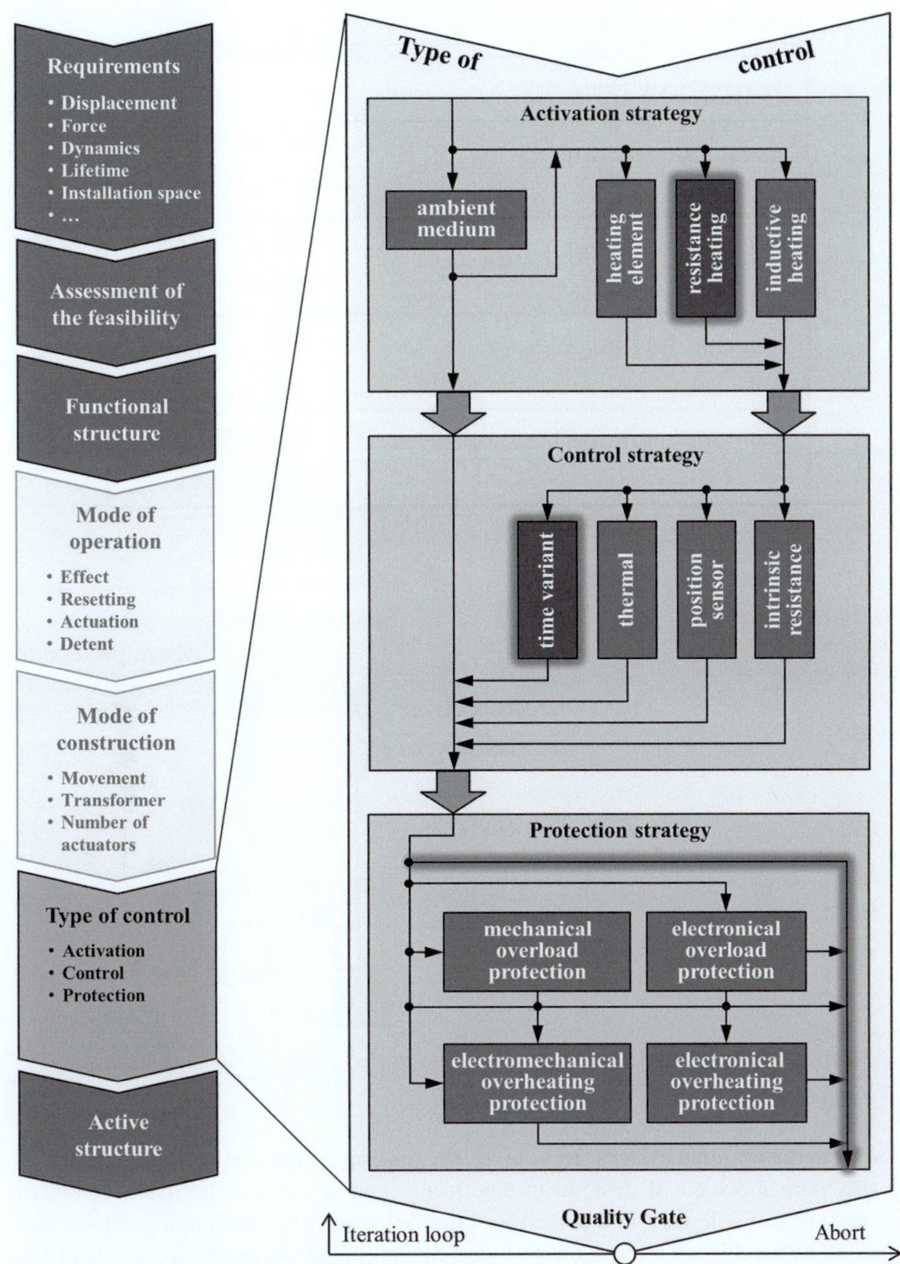

Fig. 10.12 Methodical system design – determining type of control

Table 10.9 Activation options for SMA actuators

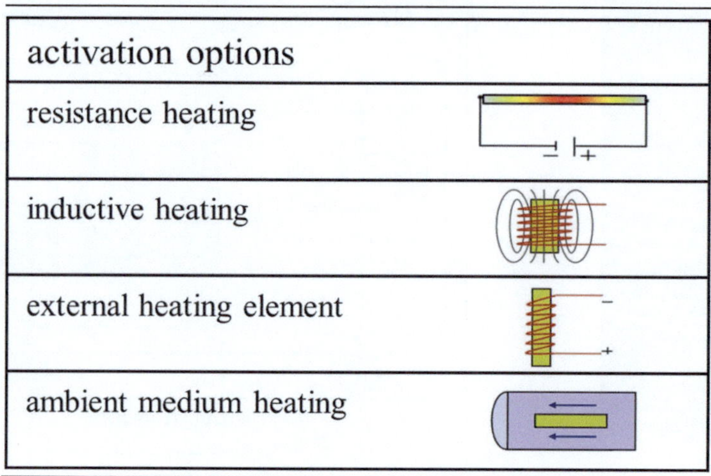

Fig. 10.13 Concept of mechanical protection against mechanical overstraining

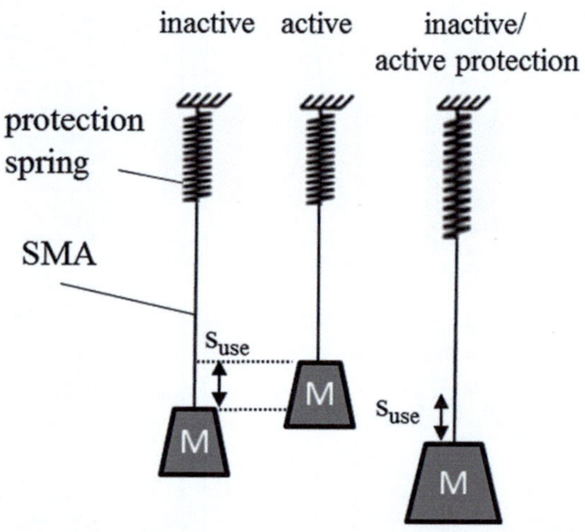

load than the nominal system stress level, the tension spring is stretched. In addition to this pure mechanical design, an electronic load sensor can be used for signal generation to implement in a controller unit. Alternatively, overloading can be recognized by intelligent interpreter software by analyzing the resistance levels of the SMA element, as discussed in Chap. 5, by stress-dependent resistance experiments. Neither mechanical nor electronic overload protectors need be implemented in the example pinch valve design. However, an electronic protection system using the resistance signal as feedback for position control or overloading can be implemented in the development of upgrades.

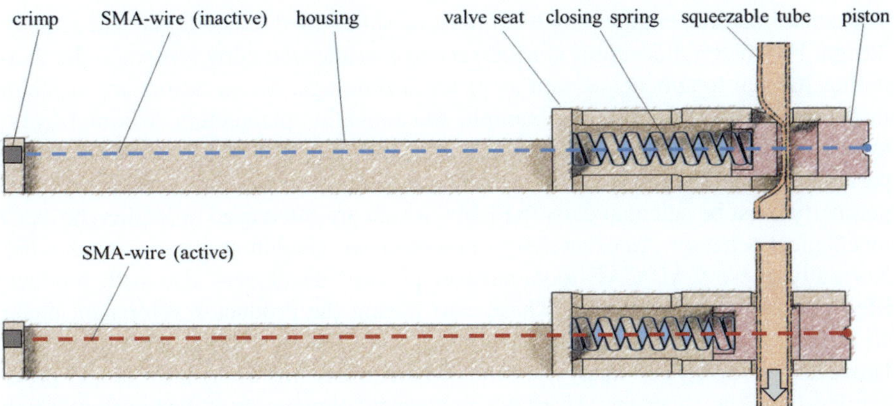

Fig. 10.14 Sketch of SMA-based pinch valve

Even overheating of the shape memory element can lead to accelerated fatigue and should therefore be avoided. Two options are generally available here. On the one hand, the actuation can be monitored. In this case, the easiest way to do this is by using an electromechanical limit switch. On the other hand, the phase transformation can be monitored. For this purpose the resistance can be analyzed. In this case, an electronic system is needed.

10.4.6 Step 6: Active Structure and Solution Concept

The previously selected active principles are combined in the sixth step in an active structure. The active structure is shown here in a concept sketch. The sketch drawing of the SMA-based pinch valve is illustrated in Fig. 10.14. By contracting through direct electrical activation, the closing compression spring is compressed by the piston, which is directly connectd to the SMA wire. The squeezable tube is opened by the piston, too, and upon cooling of the SMA wire, the closing spring deactivates the material flow.

10.5 Domain-Based Design of SMA-Based Pinch Valve

The methods used during the domain-based design of SMA systems vary widely and depend on various application patterns. With respect to Fig. 10.5, the domain-based design must be seen, in contrast to conventional mechatronic systems, as a four-domain design. The smart material engineering must be added, in contrast to mechatronic systems, which are based on available electromotors and solenoids.

This smart material engineering itself must be subdivided into material and actuator design. For material design it is necessary to consider the alloy types and the thermomechanical treatments as well as material trainings. These factors are of major importance with respect to, for example, the activation parameters, maximal working capacity, and phase transformation temperatures. The SMA actuator-relevant parameters must be considered in the actuator design. For example, a complex SMA geometry must be calculated using FEMs, which are often used to predict the exact mechanical behavior. An exact description of these calculations can be found in [6]. Normally, these CAD/CAE tools support product developers also with product-lifecycle-management criteria. These start during the product development phase with iterations and system modifications as well as modeling across domains. However, a simple FEM analysis of the SMA could be insufficient for proper product development, so an analysis of the mechanical parameters of the housing during activation is vital to assure that the generated tensile strain will deform a resetting element and not the housing, for example. This is implemented in domain-based design as mechanical design. Of course, conventional mechanical design, like 3D CAD modeling of all parts, must also be done in this step.

10.5.1 Design of Material Mechanics

Material design need not be carried out by product developers for the most electrically activated applications. Various wire manufacturers have already optimized and trained wires for actuator applications in programs. These wires are designed for long service lives and have also partly been certified according to the DIN ISO standard. Only for special applications or special material configurations is it necessary to select the alloy and the heat treatment. For example, a SMA element must have a low hysteresis for thermostatic valves. To reduce the transformation hysteresis to 1–2 K, the use of R-phase elements is necessary. To set the R-phase, however, the material must be heat-treated in a specific manner. For this, defining the material is crucial. In this chapter, material design will not be further investigated.

10.5.2 Design of Actuator Mechanics

In this phase, tests must be conducted to evaluate the design parameters. However, these attempts only relate to the actuator element or individual subsystems and not the entire system. The complete system will be tested only after the system integration phase. An experimental test for the pinch valve actuator wire reveals that the required electrical current for 1.0 s activation, using 0.4 mm SMA wire diameter, would not be sufficient to reach the A_F temperature. In Fig. 10.15 results from the numerical simulation as described in Sect. 9.5 shows that at the maximal current of 1.5 A, the 0.4 mm wire does not transform its phase completely. Additionally, the

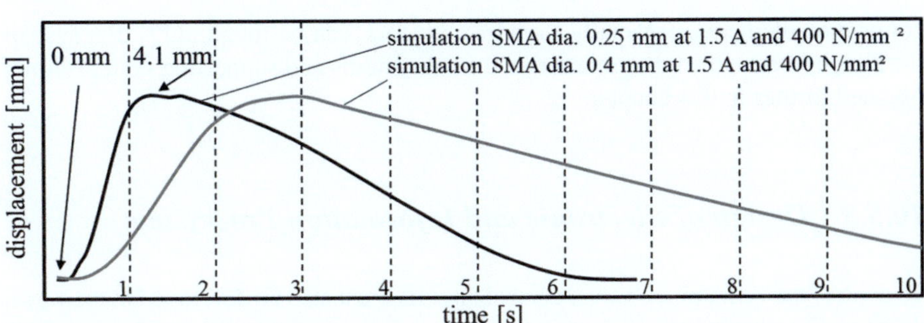

Fig. 10.15 Numerical result of 0.4 and 0.25 mm SMA wire under 400 N/mm² constant load at $T_a = 20$ °C and at electrical activation of 1.5 A

Table 10.10 Recalculated fast-track solution values for parallel arrangement of SMA wires

F_{min} (N)	F_{max} (N)	c (N/mm)	s_{use} (mm)	σ_{max} (N/mm²)	d_w (mm)	σ_{min} (N/mm²)	σ_{use} (N/mm²)
15	20	1.2	4.1	50	0.71	37.91	12.09
15	20	1.2	4.1	100	0.5	76.43	23.57
15	20	1.2	4.1	150	0.41	113.67	36.33
15	20	1.2	4.1	200	0.36	147.44	52.56
15	20	1.2	4.1	250	0.32	186.6	63.4
15	20	1.2	4.1	300	0.29	227.21	72.79
15	20	1.2	4.1	350	0.27	262.12	87.88
15	20	1.2	4.1	400	0.25	305.73	94.27

cooling speed of such a SMA wire is too low for the proper application. If the ambient temperature decreases, the needed energy input must rise to reach the proper activation level. All in all, these facts lead to the need for a system consisting of two SMA wires with smaller diameters working in parallel arrangement so as to enhance the dynamics and decrease the required electrical current level.

To recalculate a parallel arranged concept, the tensile forces can be halved (Table 10.10). This leads to significantly smaller SMA wire diameters, enabling faster dynamics and lower electrical current levels. In the example, the appropriate dynamics are reached by use of SMA wires with a diameter of 0.25 mm. As presented in Fig. 10.15, the wire generates a displacement in less than 1.5 s at the selected amperage and cools down significantly faster than the previously calculated solution with a diameter of 0.4 mm.

In addition to the design of the actuator element, important subsystems are also designed in this phase. Examples include the transformer, the return spring, locking systems, and the housing. However, support systems, such as the connection technology, must be defined and evaluated at this stage. In particular, in pull wires the connection technology plays a special role. As described in Sect. 4.4, the connection technology has a relevant impact on fatigue behavior. A detailed description of connection elements, especially for wire actuators, can be found in [7]. Because of the size, however, the interpretation of the subsystems and supporting systems is not pursued further in this chapter.

10.5.3 Design of Electronic and Information Processing

To control the SMA pinch valve, control circuits must also be developed in their own domain. The electronics, as described in Chap. 5, must also be designed using CAD tools. The selection of the proper components for control by software or analog devices can be selected by intelligent catalog systems and implemented in the program of a microcontroller or in the circuit schematics of an electronics assembly.

Within IT development, different strategies must be used to program the controller. As described in the potentials of the SMA systems, it is possible to control a device via resistance feedback in position. Additionally, in the future, it will be possible to identify the fatigue by resistance analysis in the future. Owing to the flexibility associated with implementing a new function, which requires simply changing the software of a SMA controller, object-oriented programming should be used. Within such codes, function blocks will make it possible to diagnose a system or even activate a self-repair function.

10.6　System Integration of SMA-Based Pinch Valve

Following domain-based development, the major focus is placed on system integration. This design phase brings the various domains together again and has two different approaches. On the one hand, the integration aspects of a SMA valve in an application system can be specified. On the other hand, SMA actuator integration strategies within the SMA valve device can be specified. In this context the system integration phase has as its aim to bring together the individual results of the domain-specific design. Concretely, this means that a CAD model of the whole system needs to be created. Figure 10.16 shows a detail of a CAD model of the designed SMA-based pinch valve as an assembly model with all the necessary mechanical components implemented.

By importing numerical data it is possible to combine the result of a numerical or FEM simulation with 3D CAD drawings. This make it possible to form a visual simulation of the behavior of the SMA valve system. In the example, the numerical simulation, as presented in Sect. 9.5, computes a maximal displacement of 4.1 mm after 1.3 s at a current pulse of 1.5 A with the parameters specified in the system design (Fig. 10.17). The dimensions of the drawn SMA wire are linked to the numerical data using a time-based matrix that can be obtained from numerical simulation tools.

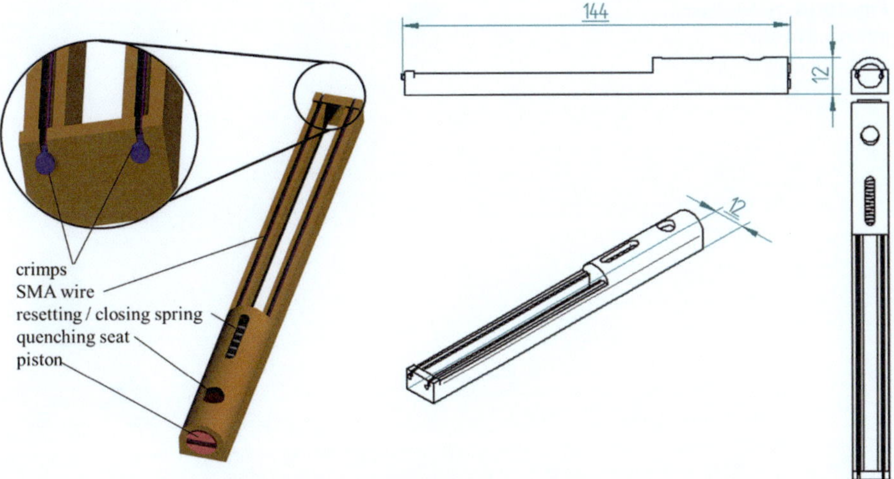

Fig. 10.16 3D CAD model of SMA-based pinch valve

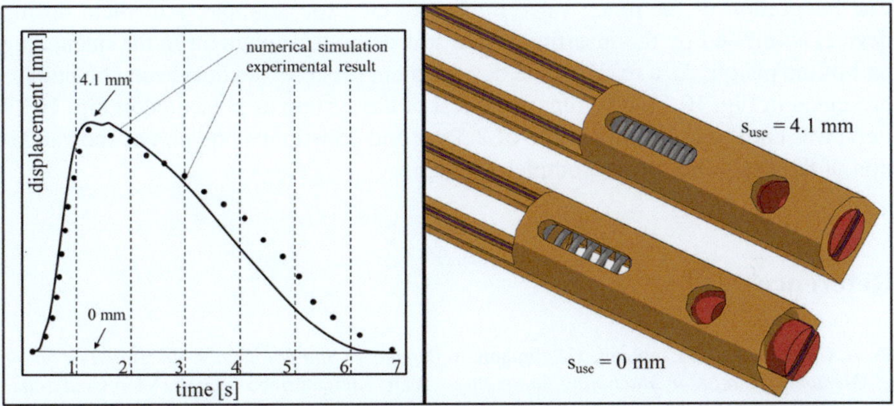

Fig. 10.17 Numerical simulation result of selected SMA configuration with experimental validation (*left*); visual simulation of SMA-based pinch valve actuation (*right*)

For comparison to a real element, Fig. 10.17 also shows the experimental results of the dynamic test of the SMA actuator. There are slight differences between the numerical simulation and the experimental results, which might derive from the system's friction, which affects the SMA element itself in the form of mechanical load. As discussed earlier, the tensional load is directly linked to the transformation temperatures of the SMA element and, therefore, with the system's dynamic.

In the example of the pinch valve, integration of the SMA wire is accomplished through the use of splices as connector elements for electrical and mechanical con-

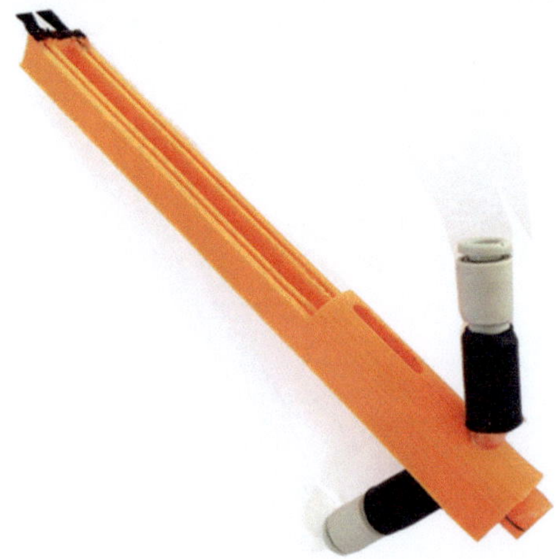

Fig. 10.18 SMA-based pinch valve after methodical development

nections. As a next step, this connection can be optimized for automated mounting. The integration of the pinch valve into a product, for example, a medical dosing device, is implied by the insertion of the squeezing tube element in the quenching part of the piston. As a result of the development process, the device can be built as presented in Fig. 10.18. The functional test of the system as presented in Fig. 10.17 (left) was carried out at a pressure of 2.5 bar and constitutes one part of the validation of the properties of the complete system.

References

1. A. Czechowicz, S. Langbein, J. Pollmann, in *Benefits of Standardization Illustrated by Shape Memory Actuators in Machining Applications*. VDE Kleinantriebe Tagung (VDE Conference on small drives) (Germany, 2013)
2. Cole-Parmer Instrument Company, Online product catalogue, http://www.coleparmer.com. Accessed 2 Feb 2015
3. VDI Guideline 2248, *Shape Memory Technology*, (Verein Deutscher Ingenieure (VDI), Düsseldorf, Germany, 2015) (in press)
4. S. Langbein, K. Lygin, T. Sadek, Significance of requirements for the implementation of new technologies using shape memory technology, in *Proceedings of the 18th International Conference on Engineering Design (ICED)* (Kopenhagen, Schweden, 2011)
5. S. Langbein, E.G. Welp, Generation of smart structures on the basis of in situ configuration of shape memory alloys, in *Advances in Science and Technology*, vol. 59 (Trans Tech Publications, Switzerland, 2008), pp 184–189
6. D.C. Lagoudas (ed.), *Shape Memory Alloys. Modeling and Engineering Applications* (Springer, New York, 2008)
7. K. Lygin, T. Sadek, S. Langbein, Development and test of force-locked connecting elements for shape memory alloy wire, in *Proceedings of the International Conference on Shape Memory and Superelastic Technology Conference (SMST)* (Pacific Grove, USA, 2010)

Chapter 11
Examples of Shape Memory Alloy Valves on Market

Alexander Czechowicz and Sven Langbein

11.1 Thermal Shape Memory Alloy Valves in Buildings and Vehicles

Especially the area of thermally activated valve actuators offers a wide and already serviced application field. Shape memory actuators are particularly suitable for a compact valve, wherein the material is thermally activated by the medium flow. In this case, SMA may combine both the sensor element and actor functionality in a single component, with no additional control is required. The phase transition can be shifted is defined by the alloy composition and by different bias voltages, so that there is a wide range of applications. For applications with a required low hysteresis may be used as already mentioned, the R-phase transformation.

11.1.1 FireChek: Heat-Activated Pneumatic Shut-Off Valve

When this safety valve (Fig. 11.1) senses excessive heat from a nearby fire, it immediately vents the pneumatic actuator and closes the air supply line. The SMA-based system triggers quickly, securing pneumatically operated process line valves. Because the valve responds to heat, not flame, it offers dramatically improved protection compared with conventional plastic tubing burn-through. It shuts off the air supply line to prevent plant air from feeding oxygen to a fire. The shape memory

A. Czechowicz (✉)
Zentrum für Angewandte Formgedächtnistechnik, Forschungsgemeinschaft Werkzeuge und Werkstoffe e.V., Papenberger Straße 49, 42859 Remscheid, Germany
e-mail: czechowicz@fgw.de

S. Langbein
FG-INNOVATION GmbH, Universitätsstr. 142, 44799 Bochum, Germany

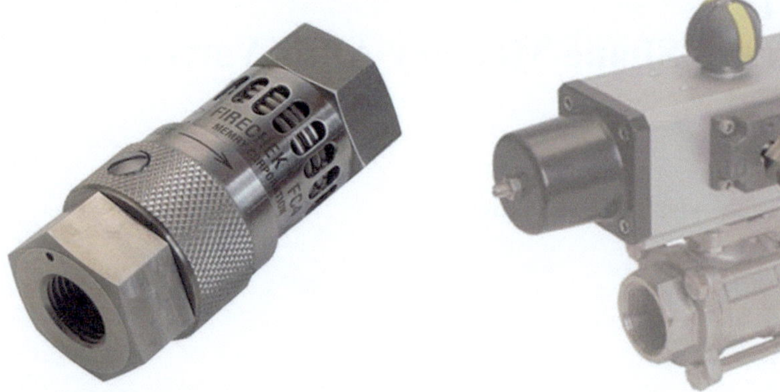

Fig. 11.1 FireChek—valve [1]

Table 11.1 Technical specifications of the valve [1]

Technical specifications	
Maximum pressure	126 PSI (8.62 bar)
Actuation time	25 s or less at a temperature rise rate of 30 °F/min
Fill and discharge capacity	1 L of air in under 1 s at 125 PSI (8.62 bar); 0.49 CV
Dimensions	2.59″ length, 1.00″ diameter
Weight	0.482 lbs
Materials of construction	Shape memory alloy actuator and sensor: MEMRY®, Thermal insulators: DELRIN®, O-rings: BUNA-N, Body cover, reset, shuttle cap, pin and bias spring: 300 stainless steel

element senses the ambient temperature and rapidly produces the force and motion to operate the valve. The manual reset of the system allows routine performance testing for safety maintenance programs. The reset is carried out without the use of tools. The valve installs easily between the pneumatic supply and spring return actuator via two female threaded connections [1]. The specifications of this valve are summarized in (Table 11.1).

Features

- Rapidly responds to temperature rise, not flames
- Vents actuator air pressure
- Closes actuator air supply line
- Needs no power source, self-activates
- Compatible with all pneumatic actuators
- No contact with process fluids
- Easy to test and reset

- Reliable, intrinsic triggering
- Easy to retrofit, installs in minutes
- FM-approved

11.1.2 SMV-Control: Valve for Underfloor Heating Regulation

Potential applications for this valve are both to be found in self-controlled cooling or heating circuits. The valve regulated the return temperatures of heating systems and the flow in fluid and gas pipes depending on temperature. The set phase transformation temperature determines the temperature of the control valve. Due to the fluid, the SMA operates at a defined temperature from self-sufficient. Such a valve for limiting the return temperature of underfloor heating is shown in Fig. 11.2 [2]. A summary of the technical specifications can be found in (Table 11.2). This valve is directly installed into the medium and consists of a regulating arbor and two opposing compression springs. One of those springs is made of a shape memory alloy. When a set temperature is reached, the spring force of the SMA spring decreases and the reset spring opens the valve seat. In panel heaters, this system regulates the return temperature independently, quickly and without external help energy around a pre-selected mean value. This avoids overheating the screed. By using this system in other fluid or gas cycles, other pathways can be connected or disconnected independently and without a bypass. The SMA system is integrated in any corresponding valve casing or can be delivered as a complete valve with common joints and sealing geometry.

Features

- Temperature-dependent regulation of flow in fluid and gas pipes.
- Use as a return temperature controller or antifreeze system for heating systems.
- Small temperature hysteresis.
- Short reaction time to temperature changes.
- Long lifetime.

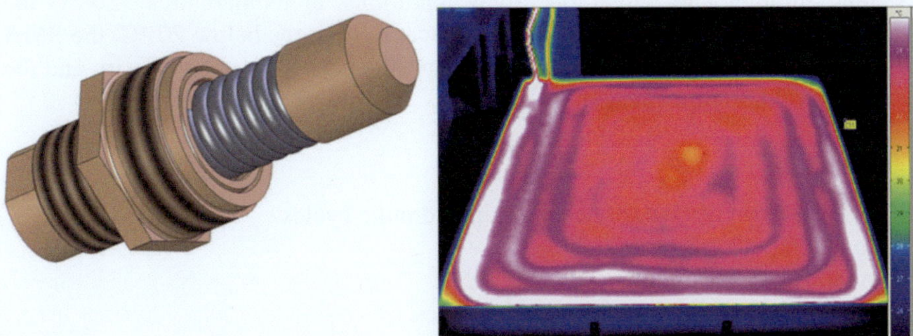

Fig. 11.2 SMV-Control—valve [2]

Table 11.2 Technical specifications of the valve [2]

Technical specifications:		
Permissible medium	Water, other fluids on request	
Permissible medium temperature	5–90 °C	
Valve-type	T40-R3	T50-R3
Start closing temperature	41 °C ± 1 °C	50 °C ± 1 °C
End closing temperature	47 °C ± 1 °C	59 °C ± 1 °C
Start opening temperature	46 °C ± 1 °C	54 °C ± 1 °C
End opening temperature	40 °C ± 1 °C	48 °C ± 1 °C
Working pressure	6 bar	
Permissible test pressure	10 bar	
Stroke	1 mm	

11.1.3 SMV-Visco: Valve for Compensation of Viscosity Changes

Using hydraulic cycles for exterior purposes, the viscosity of the fluid and the volume flow rate change due to temperature inconsistencies; this change can lead to considerable functional disturbances. Valves for compensation of viscosity differences are a solution to this problem and provide a very large application potential. Such a valve, which is not much bigger than an M8 screw, is shown in Fig. 11.3. The specifications for a valve installed in a hydraulic door closer can be found in (Table 11.3). The aim of this valve is the compensation of changing viscosity levels in hydraulic cycles caused by change of temperature. The temperature compensation valve contains a shape memory component and is capable of maintaining a certain volume flow rate within a set temperature range. The SM component is passively accessed through the ambient temperature. The depth at which the valve is placed into the casing adjusts the desired volume flow rate. To compensate for example summer and winter operating conditions, at temperatures >20 °C the valve's bypass is closed. If the ambient temperature drops below 20 °C, the SMA element opens the bypass against a return spring and adjusts the viscosity and the change in volume flow rate [3].

Features

- Compensation of viscosity changes in hydraulic cycles.
- Small temperature hysteresis.
- Small installation space.
- Short reaction time to temperature changes.
- Long lifetime.

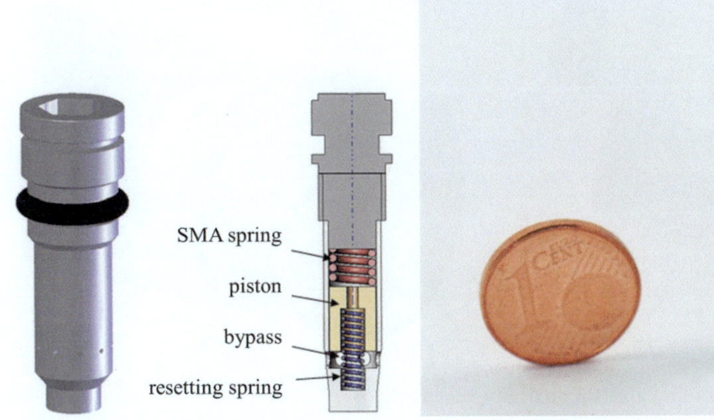

Fig. 11.3 SMV-Visco—valve [3]

Table 11.3 Technical specifications of the valve [3]

Technical specifications	
Temperature compensation	−10 °C to +20 °C
Temperature hysteresis	2 K
Lifetime	200,000 cycles
Ambient temperature	−40 °C to +80 °C
Dimensions	35 mm length, 8 mm diameter
Connection	M8x0.75

11.1.4   Thermostat Combi Valve for Auxiliary Heaters

One used in motor vehicle applications for shape memory actuators is a combi valve for auxiliary heaters. The valve is shown in Fig. 11.4 and in turn has very compact dimensions. This was also a reason for the use of the SMA in this product. The valve is presented in the right part of the figure in the martensitic state. In this state the fluid runs just trough the lesser circulation system consisting of heater and heat exchanger. The through the heater gradually heated fluid heats also the SMA element [1]. Is the phase transformation temperature reached, the SMA element begins to change shape and to move the piston [3] relative to the housing [1] against the steel spring [4]. The valve is then opened and the heating medium is now also flows through the motor circuit. Thus, the motor will only preheated when the interior is already warm. As soon as the heating medium is cooled down again, the return of the piston as a result of the restoring force of the coil spring and the valve closes again takes place [3].

Features

- Temperature-dependent regulation of water flow in auxiliary heater cycles.
- Small installation space.

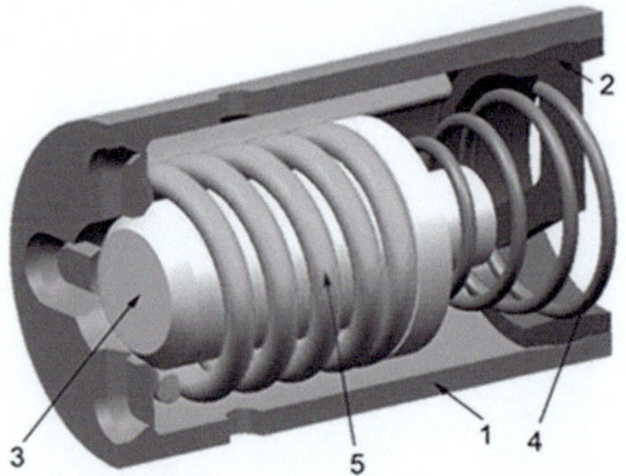

Fig. 11.4 Thermostat combi valve for auxiliary heaters [3]

- Low weight.
- Short reaction time to temperature changes.
- Long lifetime.

11.1.5 Water Temperature Control in Mixing Faucets

In thermostatic mixing faucets, shape memory actuators are used (see Fig. 11.5) for several years. The object of such faucets is to keep the temperature of the outgoing water constant. Variations in inlet temperature or pressure of water lead to significant fluctuations of the outlet temperature in conventional faucets. For shower faucets, significant loss of comfort is the result. The shape memory spring is designed for the comfort zone in this application. This means that the regulation of the temperature takes place in the interval of 30–40 °C. Increases the inlet temperature in this temperature range, there is a phase transition in the SMA spring. This results in a restoring movement against the steel spring that reduces hot water inflow. To achieve a precise controllability, with such actuators, a hysteresis of <2 K is required. Shape memory elements achieve this requirement only by the use of the R-phase transformation. Contrary to wax elements particularly fast response times and the small installation space are benefits for the shape memory technology.

Features

- Temperature-dependent regulation of outlet temperature in thermostatic mixing faucets.
- Small installation space.
- Short reaction time to temperature changes.
- Long lifetime.
- High corrosion resistance and biocompatibility.

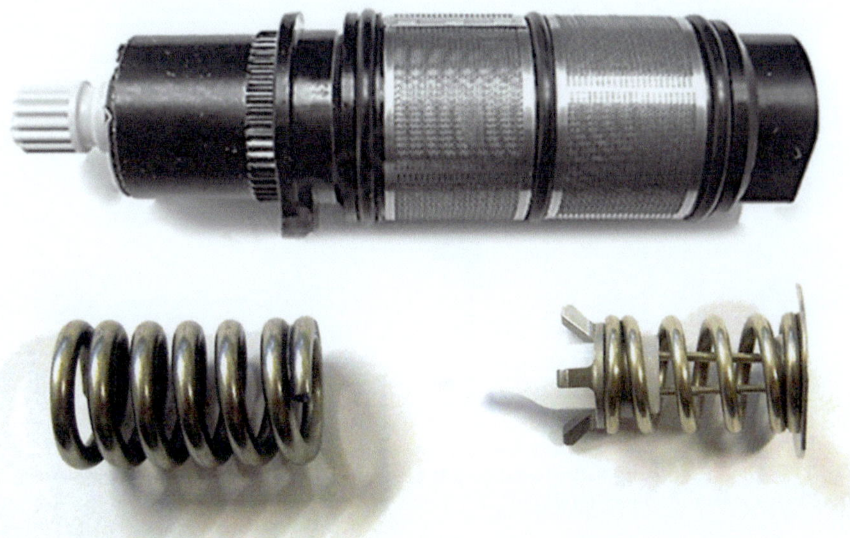

Fig. 11.5 Thermostatic cartridge (*top*) and SMA spring with R-phase effect and return spring (*bottom*) of a shower faucet of the company Toto

11.1.6 Thermal Shape Memory Alloy Valves in Household Equipment

The SMA mixing valve, build as SMA compression spring working against a conventional spring, is able to autonomously mix the temperature for the optimal realization of solubles (55–65 °C). The SMA mixer is a sensor and actuator fitted into one unique component which consists of a predefined machine set-up that does not require modification (neither the electric circuit nor the high pressure circuit). Figure 11.6 shows the application of the so-called smartea mixing unit. The pump leads directly cold water to the mixer valve, so a certain temperature level can be achieved for optimal enjoyment of tea [4].

Features

- Temperature-dependent regulation of outlet temperature of hot water.
- Small installation space.
- Precise temperature control.
- Long lifetime.
- High corrosion resistance and biocompatibility.

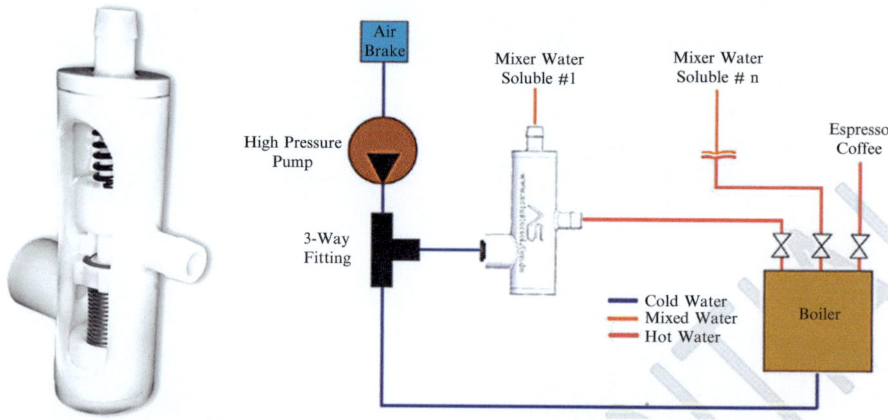

Fig. 11.6 Smartea mixing valve based on SMA compression spring (*left*), application of mixing valve in tea maker machines (*right*) [4]

11.2   Electrical Shape Memory Alloy Valves

11.2.1   *Pneumatic Valve for Lumber Support Systems in Vehicle Seats*

A successful product introduced to the market on the basis of shape memory actuators in the automotive industry is a pneumatic valve (see Fig. 11.7 and Table 11.4) to control the degree of filling of air cushion in lumbar support and massage units in seats. A low electric current activates the constrained shape memory wire and activates it to switch a flow of air through adjustment of the valve stem. The compact design is characterized by the integrated electronics, which also ensures the connectivity of the vehicle electronics to the shape memory. To meet the requirements with respect to the ambient temperature, the SMA actuator operates under a high bias stress. The pre-strain of the actuator wire is less than 1 %. Thus a long service life can be achieved. The modular design of the system also makes it possible to compile valve islands from these valve actuators and provide integrated solutions for other applications. The lightweight construction potential of this valve is considered to be very high due to the high weight saving of approximately 70 % compared to the magnetically excited control valve. This actuator is the first electronically controlled high volume application that could be enforced in the automobile. The valve has a production volume of more than five million components per year.

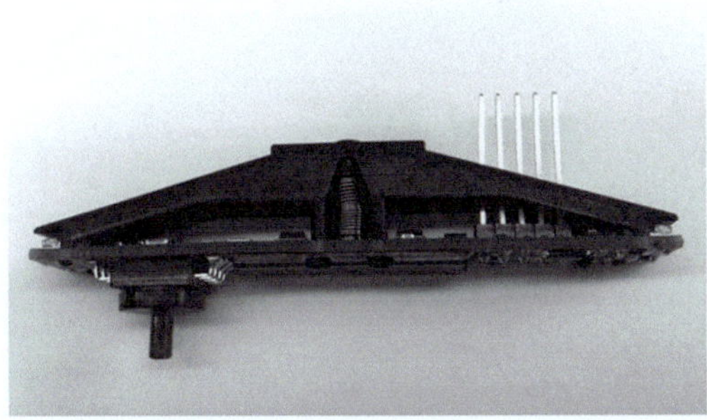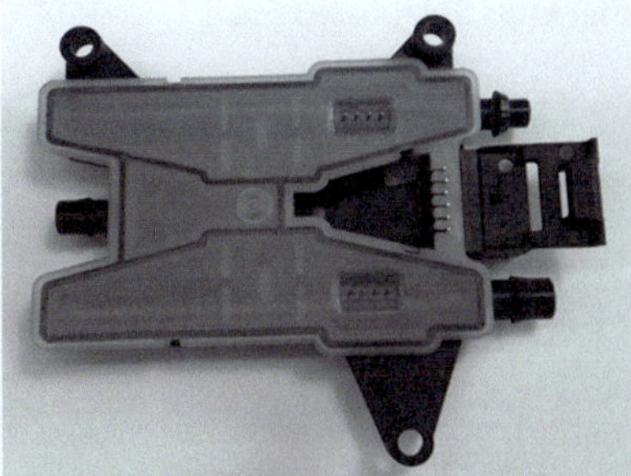

Fig. 11.7 SMA-based pneumatic valve for car seats

Table 11.4 Technical specifications of the valve [4]

Technical specifications	
Electrical voltage	9–16 V
Ambient temperature	−40 °C to +80 °C
Pressure	0–1.2 bar
Flow	<30 L/min
Leakage	<5 hPa/min
Full open/close	<120 ms
Weight	<25 g

Features

- Regulation of air flow in lumber support systems.
- Small installation space.
- Noiseless actuation.
- Low weight.

11.2.2   Small Diaphragm Valve

The micro shape memory alloy diaphragm valve (by Dolomite Microfluidics Ltd.) has small dimensions of $4 \times 16 \times 16.5$ mm comparable to an electronic relay and weight of approximately 1 g. The valve is shown in Fig. 11.8 and specified in Table 11.5. Because of these small dimensions, it is ideal for integration into microfluidic systems. It is especially suitable for low power consumption applications as it can be actuated with less than 0.3 W power. The valve is controlled using a constant current supply of 250 mA. The shape memory alloy actuation mechanism is completely silent and moves more slowly than conventional solenoid valves. The perfluoroelastomer diaphragm and PEEK body ensure that the fluid only comes into contact with inert materials [5].

Fig. 11.8 SMA-based small diaphragm valve [5]

Table 11.5 Technical specifications of the diaphragm valve [5]

Technical specifications	
Valve type	2-Way, normally closed
Orifice diameter	0.4 mm
Operating pressure range	In: 0–800 mbar, out: 0–300 mbar
Temperature range	5 °C to +40 °C
Electrical current	250 mA
Power consumption	0.3 W
Working frequency	0.5 Hz
Weight	0.9 g

11.2.3 Small Multipurpose Air Valve

The electrostem™ II valve [6], by Dynalloy inc., (see Fig. 11.9) is able to control airflow proportionally by utilization of trained SMA wires. It uses a standard Schraeder valve core or stem like those found in automobile tires. Generally, these stems are on or off depending on whether the internal stem cap is open or closed. While heated with internal resistance the SMA contracts and opens the cap; however, as it opens air begins to flow through and cool the actuator wire. Equilibrium between the electrical input and the mass of air entering the valve determines the aperture size and airflow. Consider the following test results showing how higher density of the pressurized input air restricts the output volume [6]. Table 11.6 shows the technical specifications of this valve.

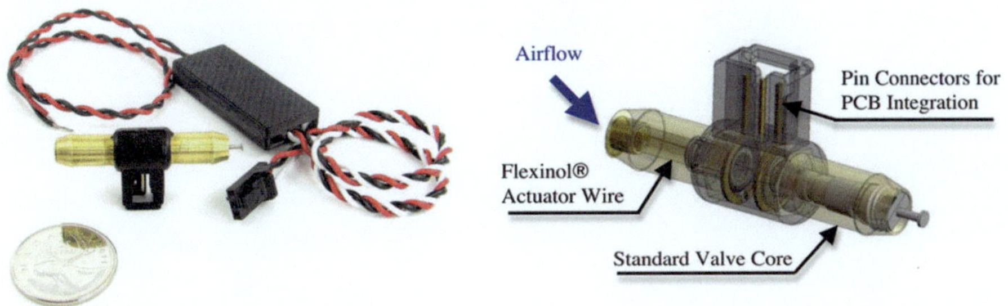

Fig. 11.9 SMA-based small multipurpose valve [6]

Table 11.6 Technical specifications of the small multipurpose valve [6]

Technical specifications	
Valve type	Normally closed valve
Operating pressure range	Up to 6.8 bar
Temperature range	5 °C to +40 °C
Electrical current	750 mA (at 22 °C)
Electrical voltage	0–5 V
Power consumption	<4 W
Resistance level	1.3 Ω

References

1. Assured Automation USA, Homepage of fire check FC-4 series valves, http://assuredautoma-tion.com/firechek/index.php. Accessed 2 Feb 2015
2. Egelhof GmbH & Co. KG, Homepage of shape memory alloy products FGA-Control, http://egelhof.com. Accessed 2 Feb 2015
3. S. Langbein, A. Czechowicz, *Konstruktionspraxis Formgedächtnistechnik* (Springer Vieweg, Mannheim, 2013). ISBN 3834819573
4. Actuatorsolutions GmbH, Homepage of shape memory alloy modular valves, http://actuatorso-lutions.com. Accessed 2 Feb 2015
5. Dolomite Microfluidics product Datasheet of Part No. 3000181, http://dolomite-microfluidics.com. Accessed 2 Feb 2015
6. Dynalloy Inc., Datasheet of ElectroStem II Valve, http://dynalloy.com. Accessed 2 Feb 2015

Chapter 12
Future Perspectives of SMA and SMA Valves

Sven Langbein and Alexander Czechowicz

12.1 Future Perspectives of Shape Memory Alloy Technology

12.1.1 Compensation of Thermal Effects by Adaptive Resetting

In order to satisfy the requirements of usability and life at high ambient temperatures, actuator systems can be designed that adapt their structure adaptive to the ambient conditions. Through this new structure, the mechanical bias of an actuator increases with the ambient temperature automatically, leading to a positive influence on the dynamics of the actuator system and on the functionality at automotive specifications. This adaptive adjustment of the mechanical bias can be realized in several variations. A schematic overview of the adaptive resetting is given in Fig. 12.1.

One of these variants is shown as an example in Fig. 12.2. Here, the adaptive reset is achieved by the substitution of the conventional return spring by a pseudoelastic SMA element. The stress/strain characteristics of pseudoelastic SMA appears as a temperature-dependent curve, which varies its plateau stress with the ambient temperature. This otherwise disadvantageous property is made here to advantage. The pseudoelastic spring element is varied, in the illustrated use case, the return spring stiffness with the change in ambient temperature. Thus increases the bias of the SMA actuator. The control range of the system is limited to the protection of the SMA actuator by a stop. The activation of the SMA actuator is purely electrically according to this concept, whereas the temperature of the pseudoelastic element can be changed either by the ambient temperature or by a controlled electric current.

S. Langbein (✉)
FG-INNOVATION GmbH, Universitätsstr. 142, 44799 Bochum, Germany
e-mail: langbein@gmx.com

A. Czechowicz
Zentrum für Angewandte Formgedächtnistechnik, Forschungsgemeinschaft Werkzeuge und Werkstoffe e.V., Papenberger Straße 49, 42859 Remscheid, Germany

© Springer International Publishing Switzerland 2015
A. Czechowicz, S. Langbein (eds.), *Shape Memory Alloy Valves*,
DOI 10.1007/978-3-319-19081-5_12

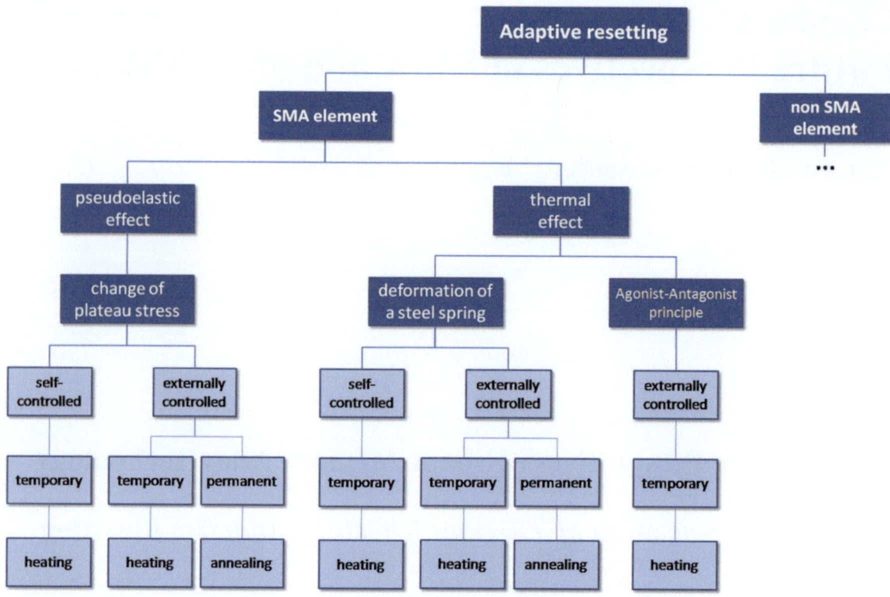

Fig. 12.1 Overview over concepts of the "adaptive resetting" strategy [1]

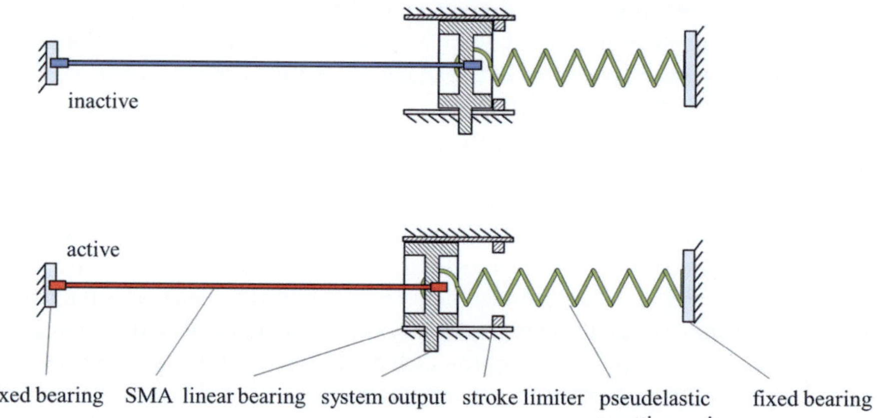

Fig. 12.2 Concept of "adaptive resetting" with a pseudoelastic element [2]

To clarify, Fig. 12.3 shows a dynamic system response of the concept shown in Fig. 12.2. The experiments presented show a substantial compensation of the increase in ambient temperature, which is reflected in an unchanged dynamic behavior.

The system behaves similarly in both ambient temperatures, in contrast to conventional designs with a steel spring. In contrast to a permanent increase in the bias, this principle has no negative effect on the fatigue behavior of the SMA actuator.

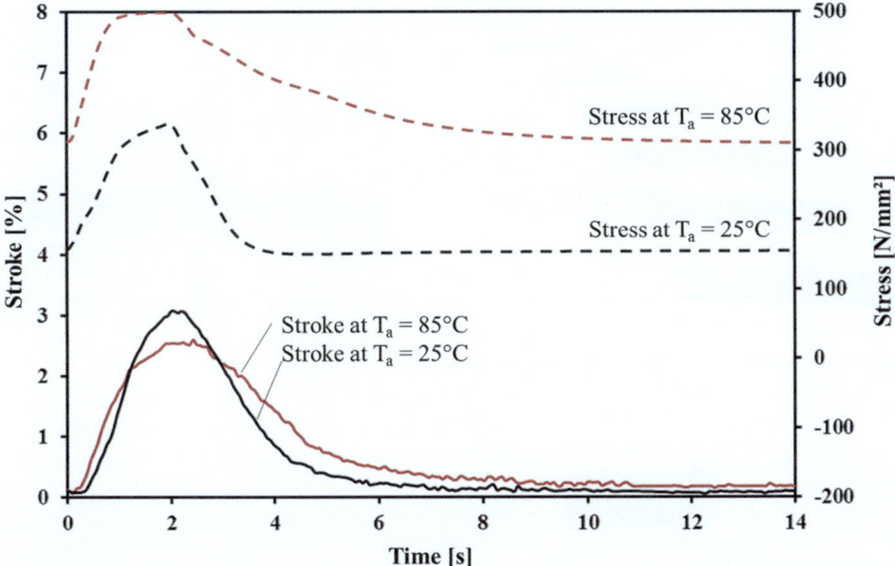

Fig. 12.3 Measurement results of an adaptive system with pseudoelastic recovery at various ambient temperatures (*black curve*: 25 °C, *red curve*: 85 °C) [2]

12.1.1.1   Example of an Actuator System

Figure 12.4 shows a standardized and fatigue optimized shape memory actuator for automotive applications. In order to enhance the actuator possibilities, it was designed as an adaptive resetting device by usage of an integrated superelastic wire. This design has mainly two advantages. Firstly, by the elongation of the superelastic wire, the resetting force stays nearly constant. Therefore, the potential for moving a mass is not decreased by the internal actuator's components. Secondly, by changing the ambient temperatures the reset force changes automatically. Thus the actuator fulfills the automotive specifications in terms of the ambient temperature. For the interior of a vehicle, the upper temperature limit is usually at 85 °C. The actuator operates at 3 A and nearly 1 V. The actuating force is approximately 4 N.

12.1.2   Compensation of Functional Fatigue by Refresh Annealing

In addition to optimizing the dynamic performance, the optimization of the functional fatigue plays an important role in the application of SMA actuators. Hence, concepts are generated to create a system that repairs itself. Figure 12.5 gives an overview over concepts of this type. In this case, one can generally distinguish between material-based and construction-based concepts. The idea of

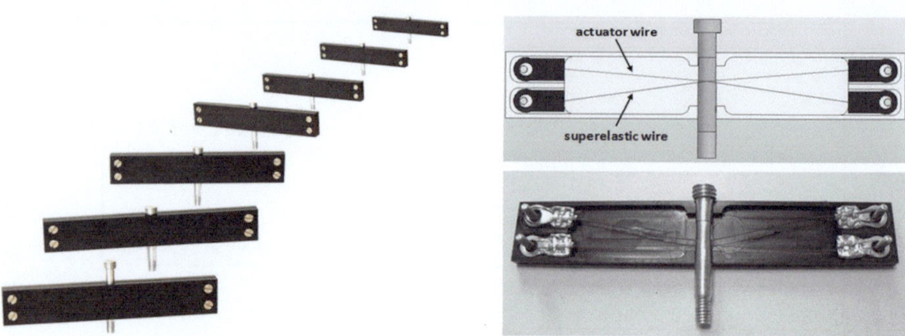

Fig. 12.4 Standardized SMA actuator with an adaptive resetting [3]

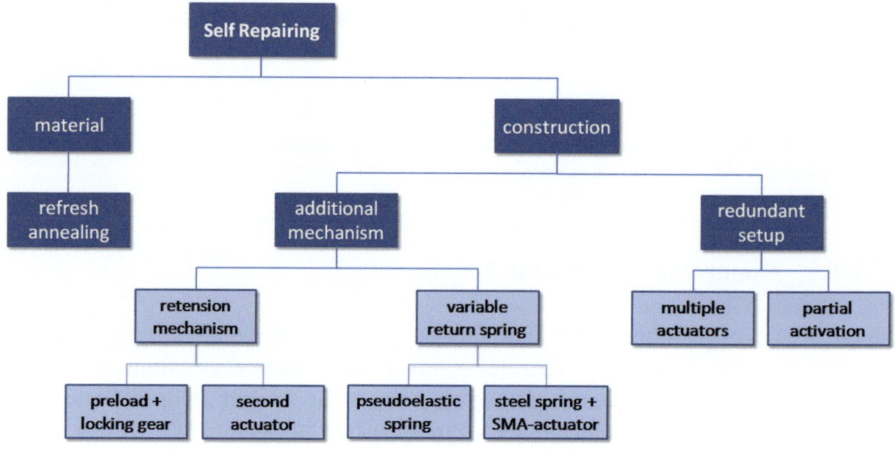

Fig. 12.5 Overview over concepts of the "self-repairing" strategy [4]

"self-repairing" pursues the aim of compensating the existing and inevitable fatigue. One advantage of the increased durability that can be achieved in this way is that the employment potential of SMA elements as drive elements in actuators can be increased substantially.

Refresh annealing constitutes one of the most important concepts within the self-repairing strategy. This concept does not require any additional mechanical components and therefore presents a simple and inexpensive possibility of a solution. This refresh process almost reaches the original quality of the material. Besides, the low annealing temperatures can ensure that a refresh process of an SMA element can take place while it is integrated. In the refresh strategy, the wire is annealed another time after reaching a determined fatigue-caused elongation, in such a way that a further drop of the stress plateau is prevented. In this process the cycle is stopped

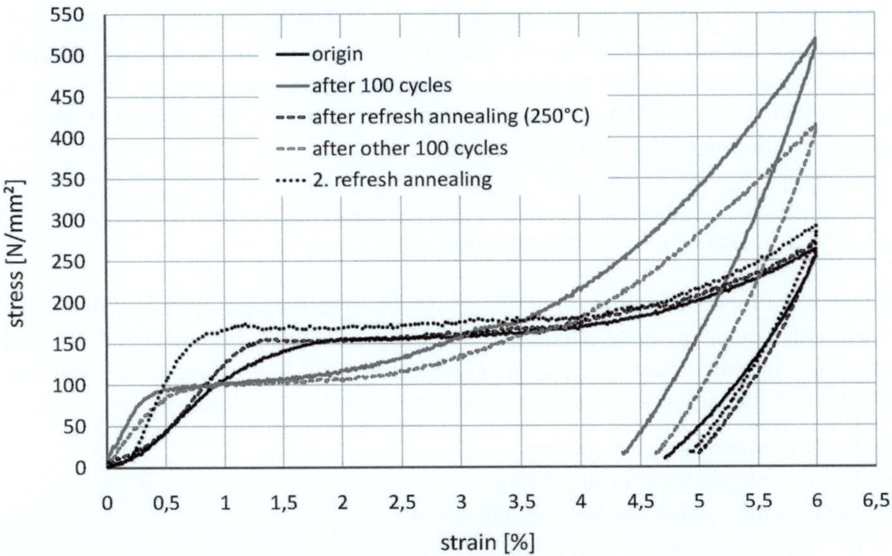

Fig. 12.6 Pulling test curves of the sample wires in different refresh phases

and the loaded force decoupled. In the stress-free assembly, an annealing process which raises the stress plateau up to a prior level can now be executed. One problem of this concept is the change of parameters like the transformation temperature, which can be seen from the longer cycle time. This problem can be tackled by a regulation of this process via resistance control [5].

The results of exemplary pulling tests of SMA wires (Fig. 12.6) of a refresh cycle clearly indicate that the changed stress plateau (here: the martensitic stress plateau) can be restored again with the help of annealing. The annealing is carried out at a temperature of 250 °C. The objects of examination are wire samples of an actuator alloy in different phases of the refresh process. The first phase is constituted by the original state of the wire. The cycling of the wire with the resulting fatigue represents the second phase. The third phase comprises the refresh annealed wire, which is cycled again in the fourth phase.

12.1.3   Functional Integrated Actuator Systems

In future, research will concentrate on continuously enhancements of the integration potential of actuator systems. In this case an outstanding feature of shape memory alloys (SMAs) is their potential to produce different functional effects like thermal shape memory or superelasticity in one component. The aim of the current research is to find a way to create a universal component with properties adjustable for various applications solely by modifying the local material properties.

This process is called local configuration. Local configuration opens up a new perspective namely the profoundly integral design to create an actuator element. Local configuration means to modify the material properties and the structure in certain regions of an SMA component. The modification of the material properties causes a variation of the SMA properties.

12.1.3.1 Basics of Local Configuration

The basis of local configuration of the material properties is generated by first deactivating the shape memory effect in the whole element and then local activation of the shape memory effect by use of local heat treatment. In addition to the local heat treatment, there are other methods to local configuration, such as:

- Local configuration via coating.
- Local configuration via structuring.
- Local configuration via alloy composition.
- Local configuration via ion implementation.

Local Configuration Via Heat Treatment

Local configuration of SMA effects via heat treatment is a method that changes the material properties on a crystalline level. This allows the realization of different functions by changing the micro-structure. The origin of this method is the rejection and/or influencing of the SMA effect in either one region of the component or the entire component. These are possible solutions for the rejection and/or influencing of the SMA effect:

- *Cold-forming* of metal sheets or wires to generate a high dislocation density.
- *Amorphous layer precipitation* during sputtering or amorphous material solidification during, melt-spinning process.
- Variation of the concentration of Ni_4Ti_3-*precipitation particles* in nickel-rich NiTi shape memory alloys.

In order to generate the desired SMA effect, the SMA components must be heat treated after the effect rejection. The setting of the shape memory effect takes place by reducing the dislocation density. The heat treatment by means of a resistance heating element represents a viable solution. It is not mandatory that the configuration of the component is made during production process. In case of the heating due to inherent resistance it can also be done at start-up or during operating process.

Local Configuration Via Coating

In case of thin-layer composition components the thermal or mechanical effect can be produced by local crystallization of the amorphous precipitated SMA layer and/or by a locally limited coating. The main method that is realized to design SMA layers is the sputtering method, also known as cathode sputtering. The sputtering method is part of the PVD (Physical Vapor Deposition) methods. In case of the sputtering method, plasma (coming from the sputtered alloy) is ignited, between the substrates to be coated and the sputter cathode. The electrical voltage (responsible for the plasma) accelerates the ions toward the sputter cathode (target from SMA). In this case atoms are ionized and deposit on the opposite substrate [6].

Local Configuration Via Structuring

The local configuration by structuring indicates the local cancellation of the coherence of the structure elements and works mostly with the separation processes. This procedure is irreversible and it also does not influence the SMA effect. These methods also have the advantage that they are realizable without high technical effort on the users' side.

Local Configuration Via Alloy Composition

Different alloy compositions of NiTi allow both the influence of the SMA properties and also a variation of mechanical values by changing their binary composition or adding ternary elements. One possibility is the local modification of the alloy composition by means of powder metallurgy. Additionally, density and hardness of the alloy can be modified by means of the particle size of the powder [7]. In the future, selective laser melting will play an important role in the production of SMA components from powder.

12.1.3.2 Local Configuration of Actuator Elements

Example of a Locally Coated Thin-Layer Actuator

A possibility of local configuration of SMA components is the local coating of SMA thin-layer actuators. Figure 12.7 shows the states of a partially coated thin-layer actuator. The SMA multi-layer film actuator consists of a substrate which is coated on both sides. A 15 µm thick molybdenum film is used as a substrate. This film was alternatively coated with NiTiCu and steel. The SMA coating was approx. 6 µm thick, the steel coating approx. 1 µm thick. The coatings were precipitated from the substrate by means of the sputtering method. The final martensitic form of the actuator film was caused by an annealing process in coiled condition [6].

The reasons for the application of a molybdenum substrate are the good adhesion of the NiTi layer and the diffusion inactivity of molybdenum. The SMA material of the actuator layer is a TiNiCu layer. Copper-containing alloys have both a lower hysteresis and a better adhesion [6]. The stress condition of the multi-layer film and therefore the deflection caused by different expansion coefficients is unequal zero, because of crystallization annealing after the precipitation.

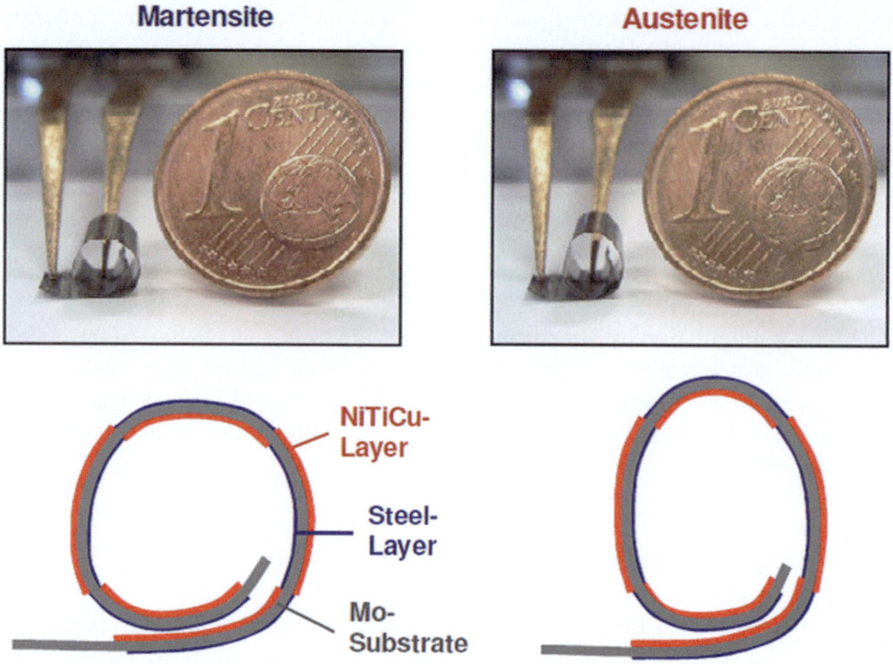

Fig. 12.7 Locally coated thin-layer structure [6]

At a stationary crystallization temperature this undesired deflection condition can only be achieved by application of an additional coating that has no SMA effect. As a result the investigated thin layer structures were coated with a steel layer.

The maximum deflection of the round thin-layer structure (diameter 4 mm) is 1.04 mm, measured by laser triangulation method. A corresponding force of 0.1 N was measured. The measurements show that these partially coated thin-layer elements are suitable in most of the technical applications. Moreover, this exemplary structure clarifies the advantage of a local coating for the generation of high forces and displacements. This behavior is not applicable for single-coated thin-layer structures.

Examples of a Locally Heat-Treated Actuators

The evidence of a cascaded method of operation was provided on the wire of an actuator alloy. Figure 12.8 shows the configuration behavior for different heat treatment temperatures. The basis for this test was a cold-deformed wire with a high dislocation density and no shape memory effect. Furthermore, as a basis for comparison, those gradients of the tests were shown having the same annealing parameter during the individual tests as the configured regions of the step-actuator. Here you will see that the step-actuator gradient can be derived from these graphs.

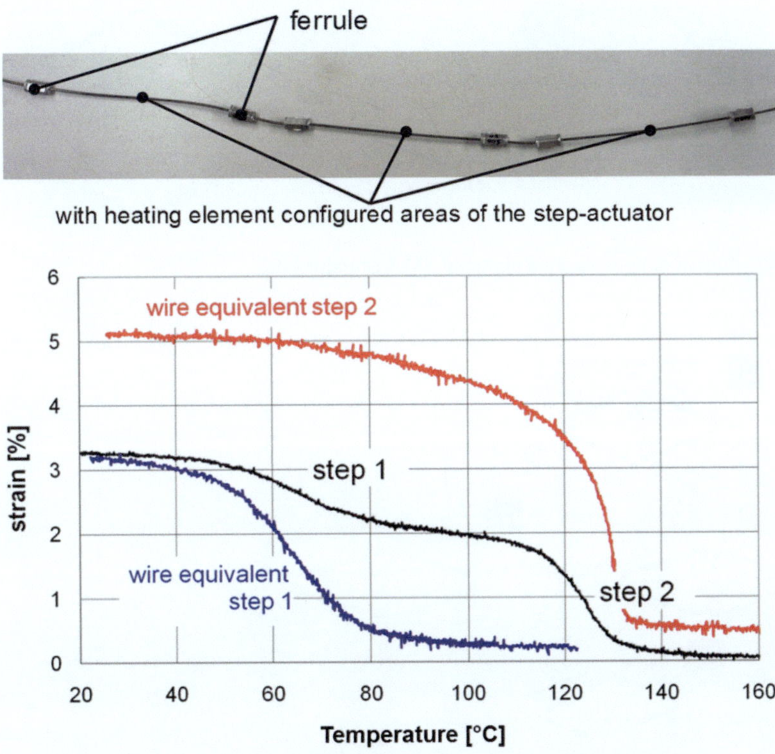

Fig. 12.8 Step-actuator based on locally configured wires of an actuator alloy [8]

12.1.3.3 Functional Integrated Actuator

To reach a simple structured actuator, this example would concentrate on increasing integration potential. While doing so, the distinguishing characteristic of an SMA, i.e. to be able to apply various effect characteristics (pseudoelasticity and thermal shape memory) in one only component is utilized. Thus, the following functions can be realized in one only SMA component: actuator function, reset function, hinges function, dampening function, and structural function. Instead, to achieve a specific local function, the elements are subjected to local heat treatment carried out by a local resistive heating element. An example of an integrated structure is shown as a further development of the actuator system from Sect. 4.5.3 in Fig. 12.9.

In addition to the above-mentioned functions, this kind of actuator also includes the function of a force–displacement conversion. Based on cold-rolled sheet metal semi-finished products, not only laser cutting but also punching methods could be applied as manufacturing methods.

                                                      S. Langbein and A. Czechowicz

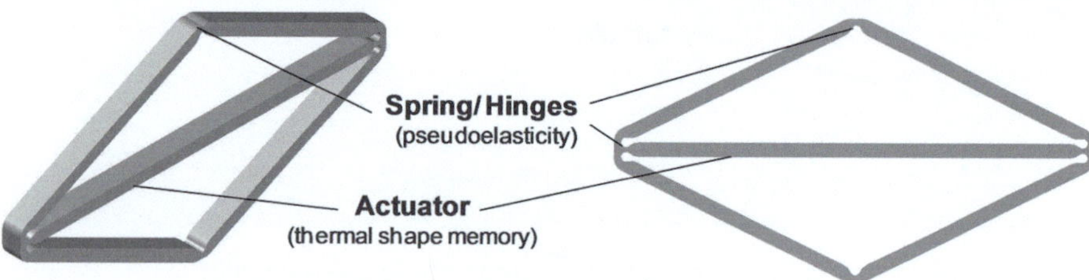

Fig. 12.9 Functionally integrated diamond SMA actuator [10]

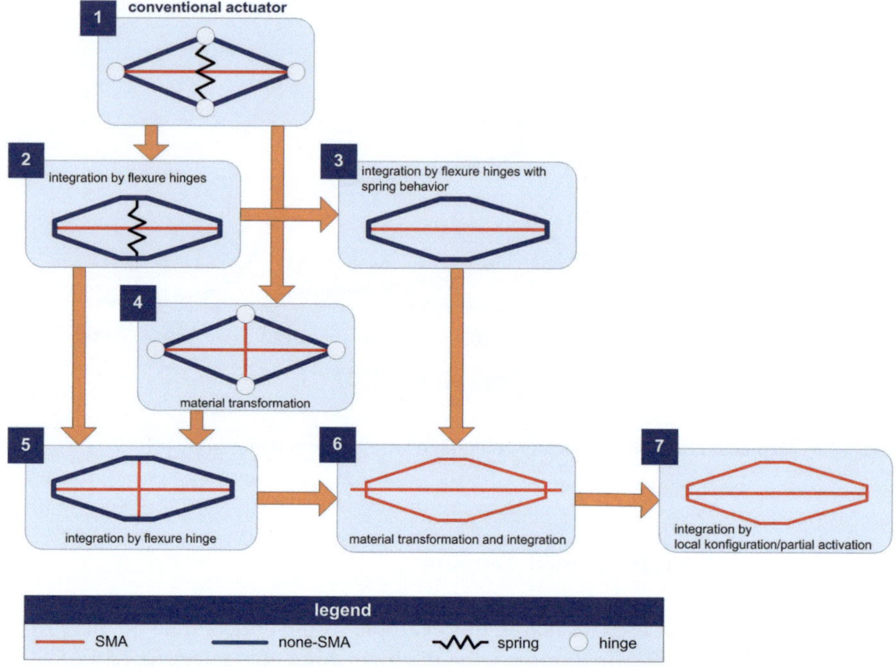

Fig. 12.10 Evolution of the diamond actuator within the design space [9]

12.1.3.4 Procedure for Functional Integration

Figure 12.10 shows the movements of the diamond actuator within the design space. Thus it appears that the aim of total integration can be achieved following different approaches. On the one hand there is the possibility to integrate all non-SMA components in horizontal direction (1 over 2–3). After that, a transfer into an SMA component using vertical direction (3–6) is performed. Finally, the SMA component can be integrated with an SMA actuator element (6–7). On the other hand, there is the possibility to start with the transformation of the non-SMA components into SMA components (1–4) followed by a stepwise integration until reaching total integration (4–7). In between, further possibilities exist where material transformation is performed at early phases of partial integration.

12.1.4 Introduction in Sensing Effects of Pseudoelastic SMA

After reaching the austenite finish temperature, SMA shows the pseudoelastic behavior at applied mechanical stress. By the constant volume deformation of the material, for example, the wire diameter of a pseudoelastic shape memory wire changes approximately linearly with strain. Firstly, the change in cross-sectional area and the change in the length itself influence the electrical resistance of a pseudoelastic SMA component. On the other hand, the electric resistance is affected by the stress-induced phase transformation of austenite to martensite. In Fig. 12.11 the resulting resistance characteristic of a pseudoelastic wire is shown. With increasing strain, the electrical resistance increases linearly. An identification of a transformation hysteresis can hardly be derived from the electric resistance signal. This is caused by an unremarkable change of the SMA element temperature.

Consequently, it is possible to derive a direct linear relationship between the strain and electrical resistance of a pseudoelastic SMA. The resistance signal can thus be used as an intrinsic sensor, which detects a mechanical deformation, or the position of a system which is prestressed by such an element.

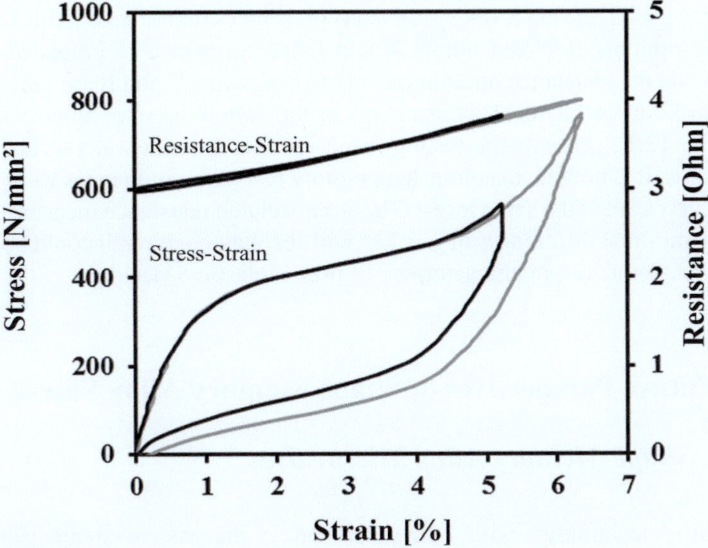

Fig. 12.11 Change of the electrical resistance of pseudoelastic SMA wire in stress–strain experiment [2]

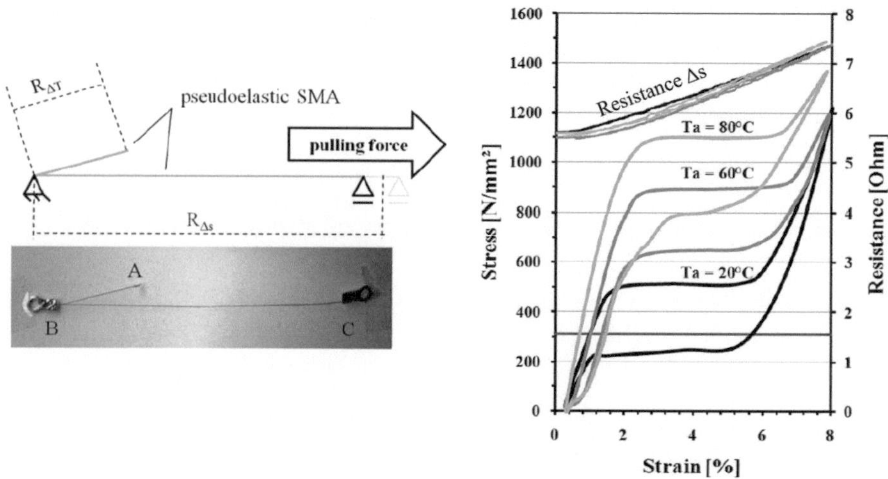

Fig. 12.12 Experimental results of testing of pseudoelastic SMA elements under varying ambient temperatures [2]

With a change in ambient temperatures, pseudoelastic SMA properties change significantly. To initiate a mechanically induced phase transformation, more mechanical energy must be initiated at increasing temperatures. Thus, the mechanical plateau stresses of pseudoelastic material increase. A change of the resistance characteristic over the material elongation is not observed according to different ambient temperatures. Figure 12.12 shows an experimental result: In the left part of the figure, a pseudoelastic SMA specimen is presented. It has two notable resistance measuring options: Between points A and B the resistance of unloaded SMA is logged, while the resistance measurement between points C and B are influenced by temperature and mechanical deformation in the following experiment. The right part of Fig. 12.12 presents the testing results of this specimen at varying ambient temperatures. It is notable that none temperature resistance influences were detected (all $R\Delta_T$ curves are at the same level). The strain-related resistance measuring curves show only minimal differences in gradient and absolute position. It could be effected by the stress–temperature characteristic of pseudoelastic SMA.

12.2 Future Perspectives of Shape Memory Alloy Valves

12.2.1 Shape Memory Alloy Microvalves

A promising technology with rising demand is the microsystems technology. MEMS, microelectromechanical systems, are used more and more in intelligent applications mainly in medical and high-tech industry. Such intelligent solutions

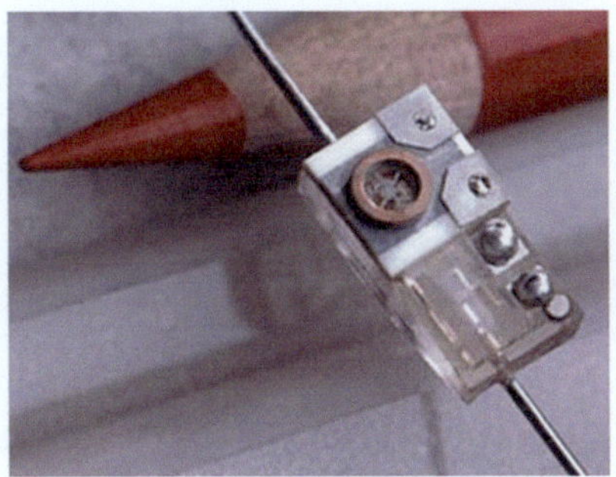

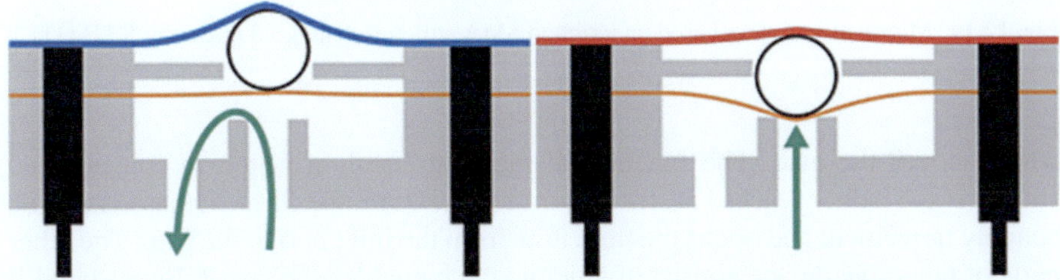

Fig. 12.13 Microvalve system with flow sensor (*upper*), working principle of microvalve by KIT (*lower*) [10]

demand usage of multifunctional materials such as SMA. An application with a high potential for the future, a shape memory alloy microvalve is presented in Fig. 12.13 [10].

The microvalve developed by the Karlsruhe Institute of Technology (KIT, Germany) consists of a polymer housing with integrated valve chamber, a membrane and a shape memory microactuator. The microactuator is deflected in its center by a spacer, e.g., a microball. In power-off condition, the microvalve is open (state 1). In this case, the fluid can exhaust depending on the supply pressure. Upon electrical heating, the microactuator transforms into its planar memory shape and thereby closes the valve (state 2) [10].

The testes gas flows reach even to 800 cm^3/min at high pressure differences up to 5 bar. Due to thermal decoupling by the membrane and spacer, temperature-sensitive fluids can be handled. The microvalve system in Fig. 12.13 has a diameter of the valve chamber of 2 mm and consists of the valve drive itself and a flow sensor. It can be used for dosing applications in fuel cell technology or bioanalytics.

Another promising concept for an SMA microvalves has been developed by the Royal Institute of Technology in Stockholm (KTH, Sweden) and is based on an SMA bending actuator. Figure 12.14 shows the conceptual design and the functional pictures [11]. The SMA actuator is heated by a golden substrate which is able to bend with the SMA. The imprinted shape of the SMA device is a bended ribbon,

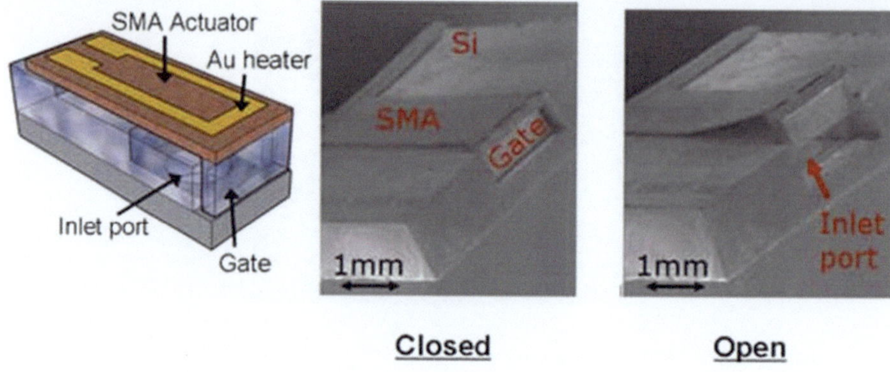

Fig. 12.14 Microvalve system based on a ribbon SMA structure with gold heater by KTH [11]

which moves the gate. The function of the gate equals a piston in conventional valves. If the heater reaches austenite start temperature, the SMA actuator starts its bending movement and opens the fluid flow from the inlet port to the gate. The other parts of the valve device consist of silicon which enables the possibility to build it on a wafer.

12.2.2 Exemplary Concepts for New SMA Valves

The futuristic ideas as presented in the upper section have to be supported by the proper system's engineering. Today's problem often is that products and services are developed separately and by different teams. Therefore, technical products are often not suitable to perceive service functions. In order to solve this problem, engineering of service-friendly systems has to be considered as presented by two different following examples for a reconfigurable SMA valve and for a customized production method for SMA valve drives.

12.2.2.1 Reconfigurable SMA Valve

In Fig. 12.15, an SMA valve modular design is showed on the left side. While the thermal valve (a) resembles generally today's SMA on-market systems, a modified component has been attached on it in (b). This secondary SMA element allows the adaptive adjustment of the system by electrical impulse or a boundary thermal field. If the valve's purpose is to control the flow of a fluid by the fluid's temperature, external thermal effects can be compensated by the secondary SMA system. This is also presented in the experimental measurement showed on the right upper side of Fig. 12.15. While the system (b) has an Af temperature of 84 °C with inactive

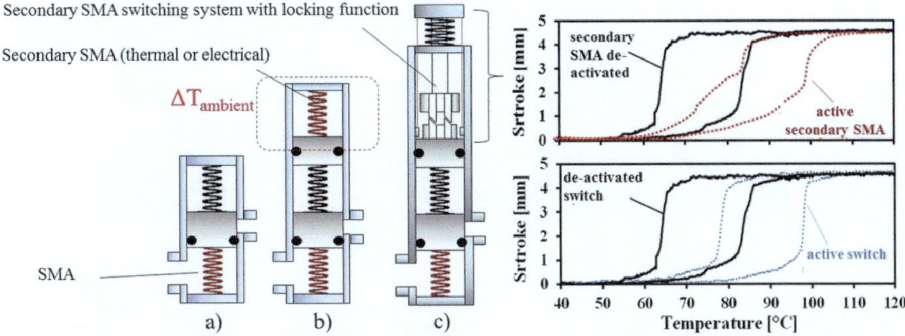

Fig. 12.15 Concepts (*left*) and experimental results (*right*) for SMA valves with attached system upgrades for enabling of modification functions in running systems [12]

secondary SMA element, the modified system reacts as well on the ambient or electrical-generated temperature. Therefore, case-sensitive adjustment of the phase transformation temperatures up to $\Delta T = 21$ °C is possible in this concept. An electric variant for the discrete modification of the valve system is showed in the left part of Fig. 4.8 in (c). An attached switching system based on electric SMA drives allows the bi-stable compression of an external spring element. This element operates as additional load on the main thermal SMA drive causing a shift of the transformation temperatures as presented in Fig. 4.8 (right side, lower part). This allows, for example, a shift of the operating point due to a change of system parameters.

Considering IPS[2], a scenario of thermal valves in machine systems is imaginable, which either compensates ambient temperatures (concept b) or a service rendered by the service supplier allowing to vary the thermal switching temperatures electrically via internet (concept b or c).

12.2.2.2   Rapid-Manufactured SMA Valve

Another important topic of IPS[2] is the product individualization. In today's markets, valves are often used in specialized applications. The opportunity for individual valves as replacement elements in thermal control systems or as additional systems (e.g. for compensation of thermal variation of fluid viscosities) can be achieved by the unconventional and compact design which is possible with SMA alloys. Therefore the combination of rapid-manufacturing technologies, such as fused deposition modeling has been combined with standardized semi-finished SMA spring elements in Fig. 12.16. The design of the system is fitted to additive manufacturing with two materials. The support structures (Acrylonitrile butadiene styrene (ABS) material) can be rinsed out in the process, while the ABS structure material is hard and dense enough so it can be used for different valve mechanics. By removing the support structures, moveable systems can be manufactured in one

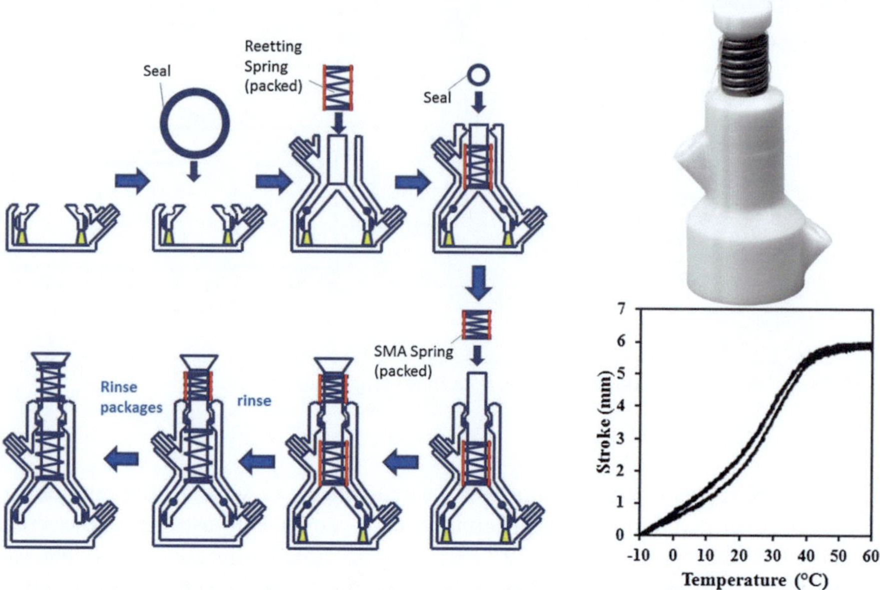

Fig. 12.16 Production of a thermal SMA valve by fused deposition production process (FDPP) with valve example and thermal testing result [12]

process. During this fused deposition production process (FDPP), seal elements, resetting springs and SMA drive springs can be inserted automatically by handling systems. It is necessary to insert the spring elements as packed elements, so they do not generate forces on the soft melted polymer parts during the process. The packing material is also an ABS-based support material in wire form which can be rinsed out during the production process. In the right part of Fig. 12.9, the "printed" valve is presented as well as the testing curve in a thermal range from −10 °C up to +60 °C. In this example, a spring with a low hysteresis was used. The SMA reacts on the fluid temperature, and opens the valve nearly linear over the fluid temperature. Following modifications of the flow diameter, the pressure and the mechanical design of the valve can be done by configurable design within three-dimensional CAD systems. Hence, product modifications can be done fast. Individual product series can be realized by the modification of the mechanical design: As function of the free installation space of both springs, the switching temperatures can be adjusted and individual thermal switching functions can be produced in this FDPP. As a result of this process, small quantities of specialized thermal valves can be produced with moderate earrings by low selling costs in comparison to conventional low-series thermal valve systems [12].

References

1. S. Langbein, A. Czechowicz, Adaptive resetting of SMA actuators. J. Intel. Mat. Syst. Struct. **23**(2), 127–134 (2012)
2. A. Czechowicz, *On the Functional Characteristics of Adaptive Resetting of Shape Memory Actuators in the Field of Automotive Applications*, in *Proceedings of SMASIS 2012 Conference on Smart Materials, Adaptive Structures and Intelligent Systems*, Stone Mountain, GA, USA, ASME, 2012
3. S. Langbein, A. Czechowicz, *Problems and Solutions for Shape Memory Actuators in Automotive Applications*, in *Proceedings of SMASIS 2012 Conference on Smart Materials, Adaptive Structures and Intelligent Systems*, Stone Mountain, GA, USA, 2012
4. S. Langbein, A. Czechowicz, H. Meier, Strategies for self-repairing shape memory alloy actuators. J. Mat. Eng. Perf. **20**(4/5), 564 (2011). Springer, New York
5. H. Meier, A. Czechowicz, C. Haberland, S. Langbein, Smart control systems for smart materials. J. Mat. Eng. Perf. **20**(2010), 559–563 (2011). Springer, New York
6. S. Langbein, A. Czechowicz, *Konstruktionspraxis Formgedächtnistechnik* (Springer Vieweg, Mannheim, 2013). ISBN 3834819573
7. C. Haberland, *Additive Verarbeitung von NiTi-Formgedächtniswerkstoffen mittels Selective Laser Melting* (*additive manufacturing of NiTi shape memory materials by selective laser melting*), Ph. D. Thesis, Ruhr-University Bochum, 2012, ISBN 978-3-8440-1522-5
8. S. Langbein, *Design of Highly Integrated Systems on the Basis of Programmed Shape Memory Alloy Components*, in *Proceedings of the ASME Conference on Smart Materials, Adaptive Structures and Intelligent Systems* (*SMASIS*), Oxnard, CA, 2009
9. S. Langbein, T. Sadek, *Design for functional integrated shape memory alloy systems*, in *International Design Conference—DESIGN 2010*, Dubrovnik, Croatia, 2010
10. J. Barth, C. Megnin, Smart materials: shape memory microactuators -microvalves and fluidic control systems, http://www.imt.kit.edu/english/399.php. Accessed 2 Feb 2015
11. W. van der Wijngaart, Gas microvalves, https://www.kth.se/en/ees/omskolan/organisation/avdelningar/mst/research/lab-on-chip/gasvalves-1.65753. Accessed 2 Feb 2015
12. A. Czechowicz, K. Lygin, S. Langbein, On the potentials of shape memory alloy valves. J. Mat. Eng. Perf. **23**(7), 2687–2695 (2014). Springer, New York

Index

© Springer International Publishing Switzerland 2015

209

A. Czechowicz, S. Langbein (eds.), *Shape Memory Alloy Valves*,
DOI 10.1007/978-3-319-19081-5